Merging City and Nature

batlleiroig

30 Commitments to Combat Climate Change

ACTAR

Index

Prologue

Exchanges

30 Commitments to Combat Climate Change

Merging City and Nature, the title of this book, is the motto we have chosen to represent our commitment to improvement in each of the actions we carry out. We work in three distinct disciplines: urban planning, landscape architecture and architecture. We seek to be highly specialized in each area while simultaneously being diverse enough to undertake any project. Today, the climate emergency is the theme that cuts across everything we do, the impetus that must guide all our actions.

Roques Blanques Cemetery in El Papiol, Barcelona (Spain, 1981)

Nature is Always the Answer

Enric Batlle

In 2002, professor and physicist Jorge Wagensberg, the first director of CosmoCaixa Barcelona, wrote a book entitled *Si la naturaleza es la respuesta, ¿cuál era la pregunta?* (If nature is the answer, what was the question?). Today, the climate emergency threatening our planet and the pandemic caused by COVID-19 have presented us with numerous questions. Perhaps nature holds many of the answers.

At Batlleiroig, we have been talking about landscape and nature since our founding in 1981, the year we won our first public competition, the landscape project for the Roques Blanques Cemetery in El Papiol, Barcelona. Since then, in one way or the other, we have remained committed to the environment and the search for solutions to address the climate emergency.

"Merging City and Nature", the title of this book, is the motto we have chosen to represent our commitment to improvement in each of the actions we carry out. We work in three distinct disciplines: urban planning, landscape architecture and architecture. We seek to be highly specialized in each area while simultaneously being diverse enough to undertake any project. Today, the climate emergency is the theme that cuts across everything we do, the impetus that must guide all our actions.

We believe it is necessary to evolve toward taking a firmer stance in all the solutions we implement, innovating in everything we do to achieve our goals differently. Some people believe the solution to the problem lies in reducing human activity. Yet they are unwilling to change their behavior and, as a result, rarely can reduce their CO_2 emissions. We believe reduction is not possible without changing the systems currently in place. To drastically reduce, we need to change the way we do things. You do not reduce by reducing; you reduce by changing, and you only change by innovating.

We have conducted an analysis of past projects, ones we are working on now and those we would like to develop in the future. This study has resulted in a sustainability matrix that expresses our commitments in relation to the climate emergency. It is a matrix that interlinks our three work areas –city and territory, landscape and public space, constructed space– with the ten concepts that we believe must be incorporated into our daily lives to combat the climate emergency and improve living conditions on the planet.

Our analysis began with the cool island concept, which in a way brings together all the others in its quest to design cities, public spaces and buildings that respond to the climate crisis. We concluded the study with health and well-being, which puts citizens at the center of the analysis, as users and the first to be harmed by the conditions on our planet. The intersection of this

Roques Blanques Cemetery in El Papiol, Barcelona (Spain, 2020)

Decalogue – Cool Island, Biodiversity, Water, Sustainable Mobility, Productivity, Recycling, Green Life Cycle, Near-Zero Emissions, Self-Sufficiency, and Health and Well-Being – with the three areas in which we work gives rise to the 30 commitments that are and will continue to be the backbone of our discussion as transforming agents of the city and the territory.

Given our passion for agriculture, we are dedicated to cultivating innovation. In El Garraf we learned to cultivate the recovery of biodiversity, while in Viladecans The Style Outlets we cultivate how to achieve 100% rainwater infiltration; in Sant Cugat del Vallès Central Park we cultivate reducing the temperature by 3°C, while in Vallmora Park we cultivate its insects; at the headquarters of Stradivarius, its biophilia, and in many projects currently under construction, we focus on cultivating the use of wood.

It is difficult to set a date for when we can be sure that all our projects will have achieved these goals. We could say 2030 to be in line with the oft-mentioned Agenda 2030, but we prefer to focus on the drop-by-drop approach, which we believe is more efficient: carrying out more and more projects that respond to these commitments, that become the example to be surpassed in the ones that follow.

We began 2020 by discussing the climate emergency, participating in the first Catalan climate action summit as a company representing the field of architecture. We continued with the process of preparing this book following the celebration of an internal congress in which the entire Batlleiroig team participated. Then, almost without realizing it, we found ourselves confined in our homes for months due to COVID-19.

The pandemic placed public health at the center of the discussion and people as the focal point when it comes to creating cities, public spaces and buildings. Social distancing has called into question our densely populated urban centers, their inadequate streets and the idea of the compact city model defended so fervently.

During this strange year in which we prepared the book, we have been asked on several occasions what solutions can urban planning, landscape architecture and architecture provide to the problems arising from the pandemic: how can we improve our cities, our streets, our homes, our work, education and leisure spaces?

We have responded in different ways to each question but always with a common thread that is useful for both the pandemic and the climate emergency: **“Nature is always the answer.”**

Exchanges

Joan Busquets
Enric Batlle

Clara Oloriz Sanjuán
Joan Batlle

Martha Thorne
Joan Roig

Ricardo Devesa

Planning as the Link Between City and Nature

Joan Busquets

"We have to understand the dynamics that have shaped the development of society, based on the performance of infrastructures which, in turn, created the city. The way that nature and the city can meet is through the understanding of their different logics, their geometries."

Joan Busquets. PhD in Architecture and urban planner. Martin Bucksbaum Professor in Practice of Urban Planning and Design at Harvard University's Graduate School of Design. Founder of BAU B.

Enric Batlle

"It's essential to design metropolitan strategies to guide us when addressing each of the issues under discussion: mobility, energy, housing, logistics, natural spaces, among others. We need projects that intensively address the urban reform of degraded and obsolete areas and infrastructures, as well as interventions that improve the extension and quality of green infrastructure."

Enric Batlle. PhD in Architecture and landscape architect. He has been associated with the Department of Urban and Regional Planning at the UPC since 1982, and is the director of the Master's Degree in Landscape Architecture - MBLandArch UPC and of the Batlleiroig - UPC Chair in Business.

Ricardo Devesa [RD] **I'd like to start with the title of the book *Merging Cities and Nature*. I know Joan Busquets suggested it when he invited Batlleiroig to give a lecture at Harvard University's Graduate School of Design (GSD). Why this title? Is it because now more than ever, in a post-COVID context, merging the city with nature makes more sense than before?**

Enric Batlle [EB] For many years I titled my lectures "The Garden of the Metropolis", the title of my doctoral thesis. I defined it as the integrated set of all metropolitan open spaces, from the usual public spaces in the compact city – streets, squares and parks – to spaces that recover the primordial geography of the city – sea, rivers and mountains –; from the interstitial spaces that still support agricultural activities that we should keep, to all obsolete infrastructures that we could transform. "The garden of the metropolis" is the expression of how cities can better relate to their public space, their exterior space and the geography that underlies them.

Later, we had the opportunity to collaborate with the Metropolitan Area of Barcelona on preliminary studies for the drafting of the Metropolitan Master Plan. Joan led this project in which we participated in the part related to green infrastructure. Several presentations and publications were involved, culminating in a presentation at Harvard. When asked to give my paper a title, I again proposed "The Garden of the Metropolis", but Joan suggested "Merging Public Space and Nature", which I thought very aptly reflected our work. I later changed it to "Merging City and Nature" – the title of this book – by elevating green infrastructure to the category of a system that allows us to better structure our cities, and because this fusion of city and nature extends to all our projects, both city and territorial, as well as to landscape and building. Talking about nature and sustainability is commonplace in many forums. We, however, don't want to limit the discussion to mere words which, over time, end up losing their meaning. Instead, we want to translate these goals into concrete actions.

Joan Busquets [JB] This is a recurrent theme in your work, and it's important because of the impasse we're confronted with. Society faces new challenges, and we can't afford to be oblivious to them. Perhaps the greatest difficulty in our field today is how to move from challenges to solutions, how to achieve the objectives these challenges put on the table. Because in the background are many components of the discussion – which we can call urban development, urban planning or urban culture – that are always more interested in studying the problem than in finding solutions. Today the discussion takes place in a context where the problems are serious, nearly overwhelming, but we don't hear very much about the solutions that could serve as the bridge between what was yesterday and what can be tomorrow. In this sense, I believe the work Batlleiroig is doing to find solutions – which are incomplete but aim to be systemic, and which are also limited but within the logic of natural space – is of great interest.

Another important issue you raised, Enric, is how city and nature are historically irreconcilable, mutually exclusive. This is similar to how they're portrayed in medieval paintings, where the city is depicted as closed, walled, not only because of production issues but because outside were demons. There's a clear contrast between the outside – the dangers – and the inside – civilization, humanity. The human need for protection against all kinds of harsh elements gave rise to the city. Humans also protect themselves outside the city. As a result, different relationships exist between nature and the city, but these are not angelic or immediate relationships. It's important to think under what conditions, in what way something else can be produced that is neither nature nor city. It will require a different name.

RD **Joan, in your book *Urban Grids: Handbook for Regular City Design*, the result of your lengthy research at GSD, you say that perhaps one of the most interesting places in regular urban grids is where they meet with geographical features and with nature. You point out that in these encounters we can identify special city programs that are more suitable for this fusion between the natural and the urban. Could this be one of the possibilities for merging the city with nature?**

JB Definitely. I think the important thing is to ensure that the natural element hasn't gotten out of control, and this is something Enric and his team have worked on. In the end, one of the ways that nature and the city can meet is through the understanding of their different logics, their geometries. Trees have a different logic than houses, but even so, a form of dialogue is possible. I think there's an interesting work area here: the landscape, which has always been of great interest to us.

EB As Joan said, we strive to find solutions to new challenges. These solutions may be partial, but they can be implemented immediately and will allow us to evaluate the results. The sum of many actions, however small, can also lead to a change in basic assumptions in the way we address the climate emergency. Each of us can make our own contribution, at our own scale, in relation to the potential of each project we undertake.

It's usual to devote a great deal of effort to evaluating – the obsession with data – everything related to these issues, but, on the other hand, a real change in the ways we do things is rare. We don't believe it's necessary to wait a long time to change how we think about issues and to choose the right strategies for their development. It can be something simple such as moving from collecting rainwater in a grating to a Sustainable Urban Drainage System (SUDS) or saving a stream in a new urban development – the geographical element that conventional urbanism would eliminate by replacing it with a piping system under one of the new streets – and making it the focus of the place on which all the other systems are structured, merging natural and urban logic.

RD **And also in how biodiversity can be implemented with the inclusion of vegetation in the urban structures themselves. To achieve this, for example, you've designed an insect hotel and installed terraces and green roofs in many buildings. To what extent is the dense, compact structure of the city capable of integrating this more fragmentary part of nature to ensure that an urbanized environment preserves a certain naturalness?**

EB The merging of city and nature is a challenge that cannot be addressed only from the perspective of conservationism of supposed virgin natures that can still be preserved in a place. We can create new natures, places where we can implement processes related to the flow of water, the recovery of biodiversity or urban agriculture and that will necessarily have to blend with the urban activity in our inhabited environments.

A quick glance at the book's title might lead one to believe we're demanding that everything should be green, overdoing the solutions to ensure that any place can be inundated with nature. As the son of a gardener, I'm aware of the maintenance difficulties that certain solutions entail – for example, vertical gardens. For this reason, we always opt for easy solutions and try to implement systems that are self-managing or require simple maintenance. We talk about logical sustainability, a sustainability that can be explained by constructing buildings with adequate solar protection full of terraces and landscaped roofs, or by developing public spaces that retain rainwater, have permeable paving, and are filled with native and productive vegetation.

We can't talk about renaturalization of the city while continuing to apply the same old solutions such as building totally impermeable streets that collect rainwater in drains. For the presentation of the Garraf landfill restoration project at the Venice Architecture Biennale, we chose the slogan "Farming Life" to explain the need to establish logical systems – in this case agricultural – which, through cultivation, promote the recovery of the life cycle in a place where it had utterly disappeared.

Landfill section with sealing specification required to prevent the entry of water and the escape of gas, thereby making the renaturalization of the place possible. 15th Venice Architecture Biennale 2016. Central Exhibition Pavilion.

RD **You both have long careers as professors. How do you think university education has changed in relation to the discipline of urban planning? How much have the contributions of ecology, geography and landscape architecture nourished it? Has their influence been noticeable in the way we used to think about the design of urban forms and how we think now?**

JB Without a doubt, I believe urban design has always been sensitive to these lateral disciplines. I also believe this is its function. All of us do work that we call urban studies, because we're concerned both with the study of the object and its possible solutions; that's why we're interested in it. At other times, I think, in general, schools differentiated between the subjects where theory is discussed, or the foundation of why and how things are the way they are, and how things are put into practice, which is what we usually call studios or workshops. It's assumed that theory is applied in these workshops, and we show how it can be done. It's more or less real, but the university is at an advantage because it can simulate actual situations without the commitment that reality imposes. The university doesn't have to represent simply the application. It isn't a substitute for professional discussion, but it can simulate other situations. In the field of urban planning these innovations always exist. But these new concerns emerge out of parallel fields such as ecology, sociology and economics, which accurately describe social or racial injustices. These are very important issues. But you have to ask yourself: is urban design going to be able to provide a direct solution to these problems? No, but it can help if it clearly defines guidelines for social integration. Yet the need to address this phenomenon is still pending in many cultures. There's still a lot to talk about concerning these issues, because real social insertion occurs through these elements, sometimes something as small as mixed uses or the priority of public space for all. I think what Enric pointed out is true: it's important to keep in mind that each scale allows us to solve or do certain things. I would say that, in general, the university must be sensitive to innovation and change. Sensitive not only in understanding problems but in identifying new problems and solutions. This can inspire different professional responses. I stress this because today the way our profession is managed is very important. We're in a situation where our workspace is increasingly controlled *de facto* by certain ways in which competitions are held, how commissions are organized, sometimes with procedural vices that end up greatly influencing the way we work or limiting innovation in our field. If we compare a housing project, whether subsidized or middle class, today to a project from the fifties, I think they're too similar. This is troubling. It's even more worrisome in the case of phenomena that are no longer circumstantial, such as the pandemic, which encourages us to think about the home in very different ways. We were already working from home. We had an office because it's there, at least in my case, somewhere I feel I can put my ideas in order. Teleworking tells us we should think about the house differently, as no longer separated into a day and a night space. Yet, there are other components that should be a priority. For example, a house that doesn't have a balcony, even if it's just a meter long, shouldn't be allowed. A balcony offers another relationship with the natural environment.

EB The influence of ecology, geography and landscape architecture is increasingly present in the various disciplines of our schools. In the field of urban planning, I believe it has been decisive in overcoming disciplinary constraints and approaching projects that respect ecological laws and are inspired by the geography of the place, and creating spaces with the people who will inhabit them in mind.

The need for innovation should be constant, but many factors hinder its application. On the one hand, well-established criteria are very difficult to change while, on the other, we have to contend with the undeniable fear of developing new solutions to replace the ones we already know. In some fields the advances are more evident. For example, in the design of public spaces or new office buildings, in recent years, we have dramatically changed the way we do things. This isn't the case for housing projects which, with a few exceptions, are still bound to very conservative types and systems. The case of balconies is typical. In the Barcelona area, a common regulation states that you can only build balconies that are 1.5 meters wide. We end up building houses with balconies eight meters long that have 12 m^2 of exterior surface. Wouldn't it be better to build a 4 × 3-meter balcony, which would allow us to make these 12 m^2 habitable? During the pandemic there was a lot of talk about the need to provide our homes with plenty of outdoor space, but so far no one has proposed changes in the regulations that we must adhere to. Many architects – including ourselves – have preferred to build buildings with flat façades that hide their small balconies instead of opting to break down the volume of the building, which would create more possibilities for having a relationship with the outside space.

In some towns, the mayor might express a desire for change. Yet, this interest will inevitably be hindered by all the technical hurdles that must be overcome. The winds of change are blowing, but its expensive to implement these proposals. It's our duty to try.

JB In that sense, the university as you describe it is a good place to verify these things and try out different alternatives. Doctoral theses should be more creative and less historicist. It's difficult for architects to be good historians. We need to address some of the problems about which other disciplines are asking architecture to define a clear position. Urbanism is a general subject in which everyone, to a greater or lesser extent, gives their opinion and proposes solutions. We have to understand, then, that, although the architect is responsible for the project, the transition from certain decisions or trends to their urban materialization in a building or public space isn't easy. It's true what Enric says. In the last 20 or 30 years, in the Catalan system, from the experience of Barcelona, a generalized knowledge and project experience vis-à-vis public space can be seen in Spain and the rest of Europe. Thirty years ago, municipal bodies designed public space based on the catalogs of the same old companies. Today, public space is viewed as an important component to which other value elements are added such as water, sun, durability and so on. There are many aspects here. The architect designs the urban space and is qualified to do so. Hence the central role of universities to provide content and answers to these problems. Here there is always a deficit, because at the time of the project, we tend to think of the standard, prefabricated solution, which we aren't going to repeat but which we'll adapt to each project. I think the value of the prototypical design is that once you've tested it, you can be self-critical and correct it. I think this is the way the urban project discipline can move forward. In every project you make decisions, you prioritize some issues over others, and you adjust to time, to decisions, to your budget. The project isn't only the beautiful form the architecture takes. It's also management, durability, participation. All these other issues can endow it with realism and social commitment.

EB In the book, we include the study involving several parks in Sant Cugat del Vallès, where we identified a difference of 3°C between different sectors, both in winter and summer, which influences green areas and constructed areas. This demonstrates that urban planning isn't harmless. Nor is landscaping, and every decision we make influences, to a greater or lesser extent, the end quality of our cities. We must be self-critical of everything we do so that we can learn what we can do better the next time. Every small change we make can help in the fight against climate change; the sum of many minor contributions can be decisive.

Parc Central in Sant Cugat del Vallès, Barcelona, 1989

A similar thing happened with the bike lane linking Barcelona – Cervantes Park – with Esplugues de Llobregat – Finestrelles. Its construction led to the need to extend bicycle lanes throughout all the municipalities in the region. Its implementation was a trigger for projects to be carried out in other parts of the metropolis that also sought to resolve existing discontinuities.

I would also like to mention a certain controversy that we sense in relation to the demand for biodiversity in cities. Our citizens want more green areas, greater interconnection with nature and an increase in biodiversity in their urban environments. However, I would venture to say that what they want is a pretty, soft, civilized and, in a way, inert nature. In many projects we promote green corridors that link the city with nearby natural spaces. These corridors are routes for our citizens but also for wild boar and other animals to enter the city. If we build rainwater retention basins that promote infiltration and the water remains stagnant, mosquitoes can appear. If we design a butterfly garden, we're asked if there will also be bees. Yes to sheep but no to wild boar? Yes, to butterflies but no to bees?

RD **In terms of health and well-being, which for Batlleiroig is one of the ten commandments that structure the book, how can we make this merger between what is built and what is natural beneficial without sacrificing a certain formalism in what is urban and architectural?**

Cyclable Connection Between Barcelona and Esplugues de Llobregat, 2014-2018

EB Nature, trees and, therefore, urban environments full of vegetation benefit our citizens' health. It's no longer a matter of putting trees in cities to beautify our streets or to recall lost nature. We know that trees purify the air, control CO_2 emissions and reduce the temperature. Scientific studies compare people's health according to their greater or lesser contact with trees, and their results support the inescapable need to build greener cities.

Inundating the city with nature doesn't necessarily mean we have to stop attempting to impose order on it. Many of our public park and urban environment projects have been structured around very clear systems intended to organize vegetation, water and the topography. In other cases, management is a corollary of respect for the original geography and the desire to establish green routes. You could say we're interested in reversing the order of the factors involved.

We all want a healthy, inclusive, safe, porous city, one not fragmented by uses or classes. In Barcelona there's a lot of talk about the superblock model, which has been praised by some and criticized by others. Personally, I was against the term because it implies fencing off the space, marking out the bad streets to obtain good streets within. I think it's better to strive for more hybrid models which, instead of delimiting, allow the different urban flows to coexist without conflict. Talking about green routes that link the different parts of the city seems to me more appropriate and, also, more inclusive.

Development of the Finestrelles Sector in Esplugues de Llobregat, Barcelona, 2004-2010

JB Then there is a larger scale. We can't forget that, at least in Europe, cities aren't going to grow much. In the last 30 years, while the Barcelona metropolitan population has barely increased, the consumed space has tripled. We can't forget the macro factors and large infrastructures that weigh on the future of our society. Although it isn't fashionable to talk about this, we have no choice but to understand the dynamics that, since the Middle Ages, have shaped the development of society, based on the performance of infrastructures which, in turn, created the city. Technologies to control water such as the construction of ports, manipulation of rivers, and so on. And later, the railroad, highways, the airplane. Any one of these elements introduces its own culture, and society adjusts to them. A very important modulation exists between society and technology. Today is a moment of rupture, of establishing new pacts between the classic heavy technologies and the city, whether we like it or not.

Powerful water systems give us our most beautiful parks. We're turning the internal logic of these elements on its head. Today, in times of powerful change, of demand for new horizons and systemic constellations, we must rethink how these makers of the city become breakers of the city. We recognize that major urban opportunities in many cities lie in recovering abandoned spaces from railroads, ports and other infrastructures. Think of Barcelona and its coastline, for example.

EB This larger scale Joan mentions is a determining factor in understanding the reality of cities. It's common to talk a great deal about city centers and little about their peripheries. In Barcelona, the compact city model – for example, the Eixample – is considered the ideal solution. Yet, this ignores the metropolitan reality and its manifold situations: a hybrid of independent cities that want to preserve their identity alongside the natural spaces we want to preserve, intermingled with multiple infrastructures and an endless number of obsolete industrial areas unable to meet emerging new demands.

We need a broader vision of the problems and the ability to develop projects to solve them. It's essential that we understand the reality of our cities and design metropolitan strategies to guide us when addressing each of the issues under discussion: mobility, energy, housing, logistics, natural spaces, among others. We need projects that intensively address the urban reform of degraded and obsolete areas and infrastructures, as well as interventions that improve the extension and quality of green infrastructure.

JB I believe we are faced with major challenges today. Still, I insist the only way to turn this global economy around is to rethink its logic. I like to be optimistic. I don't view these challenges as something negative. Rather, they represent an opportunity for a better city.

EB Big challenges require clear answers. Regarding energy production, which is currently being discussed in many countries, we lack a clear model to determine what paths to follow. It's very common to talk about a sustainable future in which we believe renewable energies are the solution. Yet, today we don't even know what model we're opting for, and any proposal put forward is immediately fraught with problems.

Countries like Denmark have drawn up a specific blueprint for wind energy and are committed to achieving self-sufficiency with renewables by 2030. What is our model? Solar photovoltaic? Biomass? Having a blueprint should be a priority, and its implementation will directly influence how we build the future of our cities and landscapes.

JB Here is where we say these critical and innovative projects make sense. Dissertations on these topics would help understand them better. Otherwise, the information is merely journalistic. The situation in Denmark and Norway is very different from Spain. They have other resources, a different population, a different political system. It's imperative to recognize that southern Europe has distinct resources and other factors that must be considered. We need to identify and generate discussion that can provide elements for projects, prototypical systems that can be generalized. It's an important and a colossal effort but a necessary one. In the book you mentioned earlier (*Urban Grids: Handbook for Regular City Design*), we discovered that the challenges that existed in the 19th century were even more daunting than today. When Cerdà conducted studies demonstrating the low life expectancy of working people, a clamor for more hygienic conditions followed and a solution was sought. Later, the urban expansion in Barcelona was reproduced in Italy and northern Europe and adapted to each place, depending on the amount of sunshine and other factors. These are interesting times. I applaud the initiative to put your house in order so that we can see the positive, prototypical and self-critical elements, as well as all the questions we need to ask in our field, which has always been limited to the spaces and opportunities that each project provides and, consequently, more removed from the development of theory.

RD **Batlleiroig has worked in the Barcelona metropolitan area for many years. The connectivity between different spaces has been crucial in your projects here, because there has been a strong desire to connect different areas. However, how do you aim to introduce the principles of sustainability in territories with different climate and cultural conditions than this one? For example, in Spain, and in other European regions, as rural areas undergo depopulation, solar plants and other energy or productive infrastructures are being built there with a decisive impact on the landscape.**

EB We're torn between a city that longs to be countryside and a countryside that longs to be city. On the one hand, some of us defend the idea that everything is a city and that its influence extends over a very wide territory. The phenomenon of the 100-mile city described by Deyan Sudjic is apparent in many other geographies. Colloquially it's said that Los Angeles ends where San Francisco begins. By the same token, we could say that Barcelona ends where Toulouse or Valencia begins. Perhaps Catalonia is a city of cities, where citizens make use of the whole territory, offering them its ecosystemic, productive and leisure services.

In the countryside there's a feeling of lack of services. The so-called "empty Spain" complains about the arrogance of the urban world and how we use rural land as a backyard for our needs. The pandemic resurrected the myth of the return to the bucolic world, augmented by the capabilities that technology offers for teleworking. Visions such as Koolhaas' *Countryside, A Report* bring us back to the idea of trying to merge city and countryside: a unique, nearby territory where we need to manage how we live and work, where we produce our food and the energy we need, and how we can increase the environmental quality – air, water, green – that will allow us to preserve life.

JB A few years ago, we carried out a research project at the GSD called "Continental Catalonia" on inland and coastal Catalonia. With some very useful maps, it shows how, starting in the 18th century, the population mostly moved to the coast while energy and basic production remained inland. This represents an opportunity to create wealth from which certain primary activities can benefit. We need an energy strategy that follows this model, that says the land, wind and water are here, as has been the case historically. I believe we need this type of vision. This is an academic exercise, not an operative professional proposal, but it offers a vision that goes beyond exceptional opportunism. I believe our thinking about systems has to do with geography, infrastructures and history. This is the only thing that can provide a logical response to such phenomena as the empty territory. I dare say, on the scale you suggest, I'm not so sure that given our current situation we shouldn't be talking about – and I'm a little embarrassed to say it – a planetary urbanism. If we refuse to address what's going on in India, Africa and so on – what we call the Global South –, we won't come close to proposing a viable operational solution to our concerns at the 2021 United Nations Climate Change Conference in Glasgow.

Roche Diagnostics Building Roof in Sant Cugat del Vallès, Barcelona, 2022

RD **It's true we've recently witnessed urban planning exercises on a planetary scale. As a result of the pandemic, for example, borders were closed and the supply of certain components such as masks, chips and raw materials was halted. As a result, do we need to think and plan with this planetary or meta-territorial scale in mind?**

JB And ultimately more civilized. Because at the end of the day, all this provides us with a lesson about how our highly advanced society can put the brakes on a little and reorient itself in the right direction. I think we should avoid embracing an apocalyptic vision, as we sometimes do, because society undoubtedly has improved immensely over the last 50 years. We need to take this discussion very seriously. Architects can contribute a great deal to specifying ambitious operational actions.

RD **One last question. Is it plausible to think that a landscape can be productive and beautiful at the same time? How much land do we need to feed the population of a big city, for example?**

EB Our passion for agriculture has led us to develop the idea that we're always cultivating. Sometimes we grow food to eat, but other times we cultivate simply because it's a logical, easy, low-maintenance system that seems attractive to us. In El Garraf, we learned to cultivate the recovery of lost life while in other projects we're cultivating the recovery of biodiversity through the collection and infiltration of rainwater. We can produce everywhere in the city: rooftop farms, photovoltaic panels on façades, biomass in local forests, vegetables in agricultural parks, offshore wind turbines, flower gardens that contribute to the survival of bees. Today, cities attend to nearly all their inhabitants' needs, except for food. In many places there's talk about recovering cities' food governance, promoting the cultivation of food within their own territory and nearby landscapes. Without going into how to determine what constitutes zero kilometer food or material, the novelty of the current approach lies in the fact that many of the open spaces within or adjacent to urban fabrics should have productive capacities and that developing them may be crucial for their survival and maintenance. I feel very close to the campaign slogan of the Barcelona Regional Council that promoted "Eating the Landscape" as the only reasonable solution for the preservation of our local landscapes.

Landscape as Connector Between Nature and People

Clara Oloriz Sanjuán

"The relationship between cartography, landscape and representation – that is, how the territory is seen, how it is visualized and how it is imagined – reflects ways of thinking about what's behind it and how it should be handled. We often use the idea of landscape as a way of thinking, representing and imagining the relationship between nature and society."

Clara Oloriz Sanjuán. PhD researcher and architect. Lecturer in the Master's Landscape Urbanism program at the Architectural Association, London. Co-director of Groundlab.

Joan Batlle

"We work *for* nature *with* nature. We need to distinguish between renaturalization and imitation, because it's not possible to imitate what nature took millennia to accomplish. When designing, we look to classic gardening and farming techniques to help us manage the water and green cycle and therefore be able to work with nature to reclaim these deteriorated spaces."

Joan Batlle. Architect. Associate Professor in the UPC Department of Urban and Regional Planning, he teaches Architecture Studies at ETSAB and Landscape Architecture at the UPC.

Ricardo Devesa [RD] **I'd like you first to reflect for a moment on how the notions of landscape and territory can be conceptualized. Which implements which? How can a territory become a landscape? What ideas do you have regarding these two concepts?**

Clara Oloriz Sanjuán [COS] It's an extremely complex question. For me, it's important to understand it's not a relationship between one thing and the other, but that the landscape operates within the territory. I see both as a practice. The landscape, in my opinion, is an additional instrument of the territory. It's an aesthetic instrument that has to do with how nature and society relate to the models that are used and how this relationship is imagined. In this sense, it operates from the territory, understanding the latter as a series of political technologies, to cite Stuart Elden, such as management policies, management programs, the way in which the territory is measured and taxed, how it is legalized. Within these technological policies, for me, landscape is one more.

Joan Batlle [JB] I agree completely. Just yesterday, I was writing an article about a territorial project that discusses the different landscapes within it. A bit like the Landscape Catalog here in Catalonia, which lists the types of landscapes found within a territory. Some of these landscapes are very small, because there's a special geographical element that generates a particular landscape that differs from its surroundings.

One example is El Penedès, which we can say contains different landscapes. I agree there's a concept of territory, which for me doesn't define specific features and which sometimes is confused with or related to political boundaries. Landscape has a clearer definition of concepts and, above all, represents our vision of a territory.

COS To understand territory, we mainly look to Stuart Elden, a political geographer who associates territory with the state, because of management policies. In this case, in the absence of policies at the county and regional levels, he considerably blurs borders and the way they're understood. The idea of cataloging landscapes is interesting because it's clearly the way we see them, and the division of the catalog is a stratagem for how we think of them.

Park on the undergrounded railroad in Vilafranca del Penedès, Barcelona, 2014-2016

RD **How do we map certain territories to see these landscape essences or isolate them in units? In your professional work at Batlleiroig, and at Groundlab, as well as in your teaching, how do you draw the landscape? How do you introduce all these questions of politics, territorial limits, complex biological dynamics and resources?**

JB It isn't a case of one proposal serves all. Each project has its own way of explaining itself according to the interests and objectives of the designer. There's another more theoretical part which has to do with how landscape is studied within each discipline. Landscape designers have gone from drawing in a static way, like paintings, to understanding landscape as a constantly evolving process. Today we're interested in representing the process of water drainage, the growth of vegetation and the flow of people. The drawing becomes something alive and dynamic, like the landscape. This, I believe, is the major change that landscape design technique has undergone in recent years.

COS It's an important question because the way in which something is drawn or mapped reveals the thinking behind it, how it is managed and how it is controlled. Cartography is another territorial instrument. This alliance has its origin in the idea of wanting to levy and collect taxes; all these instruments and the instrumentalization of cartography by the territory have a very specific history and geography behind them. In this sense, the relationship between cartography, landscape and representation – that is, how the territory is seen, how it is visualized and how it is imagined – reflects ways of thinking about what's behind it and how it should be handled. We often use the idea of landscape as a way of thinking, representing and imagining the relationship between nature and society. Cartography has played a very important role in how territories are represented and their aesthetics, since cartography generates an image that describes how the territory is or will be. The cartography-landscape alliance is also very interesting in terms of how to instrumentalize the technologies inherited from geographic information systems and satellite images that are being appropriated and change the way we think about landscapes.

JB As a Landscape and Architecture professor at ETSAB, I find the differences between the two degrees interesting. The way of understanding topographic plans is very different; the interests are different, because the training is different. It seems important to me that cartography, for a topographer, should not be taken as a basis for designing. Rather each designer should analyze the landscape differently, to be able to design in their own way.

COS Yes. In fact, cartographic work is a project in itself. Often when we see a map, we think of it objectively, when in fact this is a method cartography uses to impose its power dynamics. That's why it's important, as you say, to appropriate and subvert these supposedly official cartographies in order to suggest new readings.

RD **You both agree that the landscape is a living thing, not a mere static final image. In fact, Clara, at the end of your book *Landscape as Territory*, you talk about a certain landscape fetishism. Also, at Batlleiroig you have specialized in recovering badly treated territories with the aim of renaturalization. Yet to what extent can a landscape be renaturalized?**

JB I think we need to distinguish between renaturalization and imitation. Our position has always been not to imitate nature, because it's impossible. We can't do what nature took millennia to accomplish. When designing, we look to classic gardening and farming techniques to help us manage the water and green cycle and therefore be able to work with nature to reclaim these run-down spaces. Over time, nature will effect the desired renaturalization and take over the spaces as long as the previous work of control and preparation has been carried out correctly.

COS Also, imitation involves the danger of renaturalization: renaturalize to when? A hundred years ago? Five hundred years ago? A thousand years ago? A million years ago? Stuart Elden also talks about the idea of reterritorialization, of understanding the landscape almost as a palimpsest. When one inserts oneself in a project or an intervention, one becomes part of the reading of history. It remains a reinterpretation, because, as you said before, as human beings we have a particular way of thinking and interpreting.

Other dangers are the decisions concerning what's going to be renaturalized, where the naturalization will take place, who will benefit from it and who will be negatively affected. The idea of wilderness is highly problematic, as William Cronon has argued, because of the way native populations were displaced in the United States to create natural parks.

And regarding the matrix that you mention in the book, one of the questions I had is how to translate these principles and strategies into policies, since so often it's against the law to implement them. I don't know if in your experience any project has triggered a change in legislation and resulted in management policy.

JB Here the discussion with the government or managing bodies takes place on a daily basis. We encounter an abundance of regulations that try to parameterize or overcalculate values that nature has always viewed in a more logical and, above all, practical way. One example is water management. Until today, runoff or watercourses have been calculated as if they were a gas or drinking water facility. These calculations have ignored, for example, the value of a tree in the middle of a river: the tree absorbs, retains, filters and evapotranspires an amount of water that reduces the risk of flooding downstream. I also believe there's a growing predisposition on the part of government to change and accept new ways of doing things. We didn't invent these sustainability matrix concepts. Still, in Barcelona or other urban areas their radical implementation isn't easy. Another example that surprises me is the administrative requirement in many cases to talk about plant units, the search for the perfect tree I call it, when the interest should be in the community. We want a green city. Therefore, the objective must be communal and considered from the sum of the parts. Planting trees every 13 meters doesn't renaturalize a city.

COS One of the ideas we're trying to develop in the Landscape Urbanism master's program, which we coordinate at the Architectural Association School of Architecture, is that designers should step over to the administrative side. In the 1950s and 1960s in England, a lot of architects worked in public agencies, but the numbers have fallen dramatically since then. In the end, these policies and laws determine whether these principles can be implemented. This was also the case with Groundlab in a project we developed in Santiago de Chile. The project placed a lot of emphasis on pedestrians, bike lanes and public transport in the experience of the urban and civic space. Legislation then followed stipulating that cars can't circulate at less than 80 km/h. When we talk about seeing the landscape through the territory, then, we need to understand how all these policies condition the shaping of landscapes and, sometimes, how this way of seeing landscapes gradually transforms policies.

Agroparc Penedès for Ametller Origen in the Alt Penedès region, 2021-2026

JB We're currently working on the Agroparc Penedès project, which the Catalan regional government has declared to be of territorial interest as a circular agrifood cluster. Our approach has run counter to what might be considered conventional in classical urban planning. Instead of planning and then thinking about the specific project, we started with the project and worked towards defining all the details. Now we are breaking it down to determine how to process it administratively so that each administrative body can check off the box they want or the one that corresponds to them. It's a very ambitious project, where energy or consumption externalities of industry are generated and returned to the surrounding territory, closing the circle. We're changing our way of working as urban planners and landscape architects to adapt to the laws and, finally, to implement the project as we originally conceived it.

RD **Landscapes have a productive – agricultural, energetic – component and an element of economic reversal for society. How is this fact included in the discipline of classic landscape architecture? And how are these projects articulated in coherence with existing legislation?**

COS In England, agriculture management and forest management are in different departments. When agroforestry strategies are proposed, subsidies aren't available because they're outside that scope. This means we need to think about projects differently. Here the CAP (Common Agricultural Policy) policies in the European Union have, to a certain extent, shaped landscapes consisting of agricultural fields with very few trees that damage biodiversity and cause floods and natural disasters, but no one is asking how these landscapes are being developed to prevent these disasters from occurring. In productive landscapes, some techniques become territorialization devices. In some areas where there's mining or highly specialized agriculture, technology transforms a range of conditions, from labor to climate. For example, the book *Landscape as Territory* mentions a project in Navarra from a few years back, an irrigation canal that made it possible to use this water for fields of a minimum of five hectares. Thanks to satellite images, we can see, over a decade, the total transformation of land ownership, peaks in labor, flooding problems, etc. In a way, it's a device that, through infrastructure and policies, produces operational landscapes, as Neil Brenner defines them.

JB Maria Rubert, professor in the Department of Urbanism at the ETSAB, always says:

What's more urban than an infrastructure network that connects the city to the territory that supplies it? You ask an interesting question. We always see the landscape as productive. But what does productive mean? That it feeds us? That it provides us with oxygen? That it provides us with materials? We must understand the landscape as something alive and rich, not as the empty space of classic urbanism.

Forest Path in Roques Blanques Cemetery in El Papiol, Barcelona, 2017-2020

RD **We shouldn't formally imitate nature but learn from its processes. What concepts and solutions from nature have you applied to your projects, both in teaching and in practice? What knowledge can we extract from nature that can help us engage in a dialogue with the city?**

JB We like to think we work *for* nature *with* nature. Water management provides very clear examples, based on bioengineering systems, which take knowledge from natural processes, as in the Forest Path project in Roques Blanques Cemetery. Here we used a bioengineering technique, based on beaver dams, to contain the land and achieve the project's objective, which is a paradigm shift in the way a cemetery is built. When we work with landscape in non-urban areas it seems simpler, but in urban areas it's as if humanity is transformed and no longer considers these natural techniques viable. Which is when we pour concrete, when we waterproof. Education in this sense is extremely important.

Architectural designers must constantly research and develop ideas to innovate and evolve toward absolute respect for the planet. We can't rely on the status quo and avoid making changes. Caring for, improving or creating natural ecosystems is everyone's responsibility, and anyone can influence them. We must all be willing to innovate. We must overcome our fear of trying and doing new things. It's our obligation to be enthusiastic about innovation.

COS In this sense, I think it's important to understand that the city creates subjects. Often cities designed around cars produce car-dependent inhabitants. Regarding the question you were asking, I find the distinction between the natural and the artificial interesting. The mountain you visit isn't natural; nor was the city built out of nothing. Everything that surrounds us in the city is built with natural materials, from electronic devices to concrete, and it's an indication of how we relate to nature. We need to reflect on this more immediate relationship, which is disappearing in the shadow of all the consumption chains of global capitalism. There are hydrologic systems in cities, but we have routed them through pipes and made them invisible. Also, there's a lot of danger in these nature-based solutions, which are fetishized by those who have an interest in selling them. They're presented as natural or eco-friendly technologies, but no consideration is given to where the materials for these technologies or solutions come from.

We need to understand we have different interpretations of nature, and from these we produce what we understand to be nature-based solutions. We say that nature is emergent, it's sustainable, it's resilient... We must understand the artificial part behind it. Douglas Spencer explains very well how "nature is the dummy" and that we make it talk in different ways.

The concepts we use the most are ways of understanding nature as you often refer to hydrology, that is, understanding the consequences, which aren't always immediate in space and time. We also need to understand our relationships of dependence and reciprocity that are very present in indigenous thinking. In her book *Braiding Sweetgrass: Indigenous Wisdom, Scientific Knowledge and the Teachings of Plants*, Robin Wall Kimmerer talks about these reciprocal relationships with nature, about not thinking of nature as a source of resources or these models of efficiency and sustainability.

JB What you say about differentiating what is natural and what is not is interesting. For example, we can't say the concept of primordial nature exists in Spain. Here nature is all anthropized. We must relativize these concepts and understand the landscape both where there has been anthropization and where there has not, understanding the history and what has happened before we start our work. It's surprising how quickly nature has adapted where there have been natural disasters and how slow humans have been to understand what has happened.

RD **COVID has accelerated the call to return to nature. There has been a lot of talk about the biophilic city. Could you comment on this concept and explain how we can implement it in landscape projects? What benefits can it bring for people?**

JB Biophilia is the innate relationship humans have with nature. It's that magnetism we feel. A harsh example was the lockdown and the collective need we felt to go into nature. We try to translate this principle into landscape and architecture through maximizing nature in our projects. We also do this by recovering permeable and natural solutions that cities seem to have forgotten. Our sustainability matrix is the everyday tool we use to renaturalize our cities and return to human beings the innate relationship with nature that we need. It's important that landscape architects, urban planners and architects recognize this need and communicate it when designing the cities, landscapes and buildings where we live.

COS This post-COVID flight into nature reminds me of the era of industrialization when the idea of the park as a public space for play emerged, including the natural parks we mentioned before. This was the result of cities being so dirty and polluted that escaping them became a necessity. Wouldn't it make more sense to take care of the space in which you live every day instead of fleeing sporadically, if you have the means, to what you consider clean or natural? Well, that situation says a lot about our current experience of space in cities, from which we literally need to escape. For me, the presence of green spaces in the city is, in part, a matter of social justice. I know in Barcelona it's a bit different, but many studies show how the density of green spaces and number of trees vary according to urban population strata. For me, it's essential to be able to breathe clean air, to have contact with nature, no matter what part of the city you were born in. It's a public health issue.

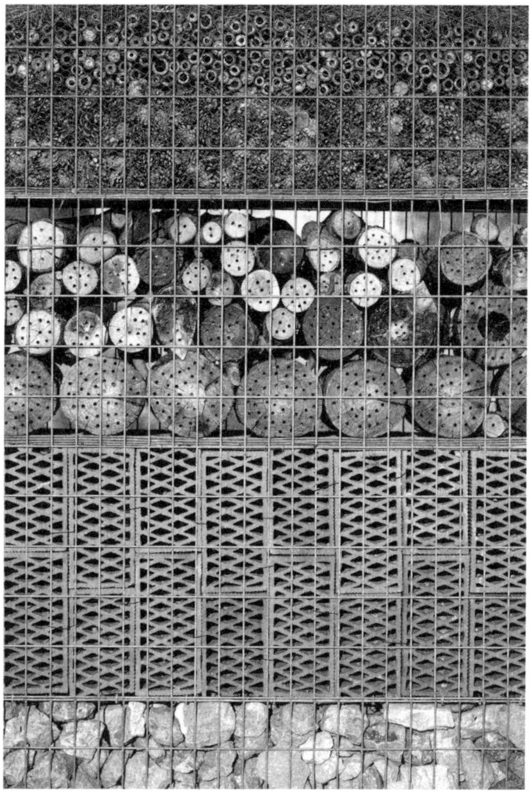

Insect Hotel for Urban Parks, 2018

RD **With the aim of questioning and expanding the role of the architect as we see**

it, Batlleiroig is exemplary. In fact, the matrix that structures this book shows that you think comprehensively about how these disciplines (urbanism, architecture and landscape architecture) ultimately merge. What do you think the new role of the architect is, as defined by these new technologies including geography, sociology and hydrology, beyond design and aesthetics? How do you think this role has exploded or imploded into a single way of seeing the world?

JB This idea of being transversal is somewhat contradictory to having departments (planning, landscape and building), but we like to think of ourselves as cross-cutting in nature. The world is becoming more specialized every day, but architects like to think that we're still calculating structures, planting trees and being urban planners. Because we have so many specialists, we're able to work across the board thanks to a team of more than 140 people with different profiles. Being transversal and innovative, we're able to take on numerous commissions and very different types of projects, always related to an area of expertise and interdisciplinarity.

The future forces us to innovate, research and develop in order to grow as a company. The planet doesn't need us; we need it. We must adapt to change. We must understand the climate emergency. But above all we must act enthusiastically to innovate – in short, to provide a comprehensive response to our profession and to society.

COS In England we work with think tanks, not only on urban planning, architecture and landscape issues but beyond. We have contacted sociologists, economists, and even dealt with legal issues. It's very important to expand, but this doesn't mean architects assume all the roles or have five heads. It means creating space for these conversations, for this interdisciplinarity, to think about what processes exist, what policies are being generated, how a designer can contribute to this discussion, what tools we have, how we operate on the basis of spatial knowledge and how we can contribute to conversations that take place at administrative levels. For example, we worked on a project to imagine Green New Deal policies in Glasgow. It was about building an image through conversations with many actors, people from local government, from organizations, from the community, with an emphasis on how people were represented. It involved thinking about who was going to be at the forefront of this image, what the ground will look like, where the materials are coming from, how it's going to be maintained, how the community is involved. In a way it was bringing all these issues out into the open rather than hiding them, as landscape culture has often traditionally done. Aware that we come from a culture with a specific historical baggage, from a critical point of view we can instrumentalize such a powerful tool as landscape development, with that aesthetic, how it's perceived, how it's experienced. I think a change in mentality is important, one that embraces Jane Hutton's view of reciprocal landscapes. It's like the idea of consequential landscapes, which isn't about understanding buildings as entities in themselves but with all their material and labor connections, all the flows mobilized in each project, their consequences and their effects.

Building as Nexus Between People and City

Martha Thorne

"Architecture must be about more than efficiency, functionality, or the response to a client's wishes. It must have the human side that distinguishes it from what technology or artificial intelligence can do. It must incorporate a message that is able to resonate with the community and society, both in the short and long term."

Martha Thorne. Bachelor of Arts and Urban Affairs and M.A. in Urban Planning. She is Dean of the IE School of Architecture and Design. Formerly the executive director of the Pritzker Architecture Prize.

Joan Roig

"The goal is to put nature at the center. Nowadays, it's inconceivable to design a building without thinking about its hydraulic, climate or biophilic behavior. In fact, this is the only way to respect the structures of the place and, therefore, relate empathetically to the environment."

Joan Roig. Architect. He has taught classes in Architectural Project Studies at ETSAB, as well as in the Master of Landscape Architecture program at the UPC. He has also served as director of BIArch (Barcelona Institute of Architecture) and cultural spokesman of COAC (Catalan Architects' Association).

Ricardo Devesa [RD] **I'd like to start by asking what nature means to you and how you understand it as an architect who incorporates it into your work.**

Joan Roig [JR]: There is nature *per se* that develops independently and has a structure that allows it to continue evolving according to its own parameters. This evolution and these behaviors aren't always good for the people who inhabit the natural environment (remember Oscar Wilde's comment about human's inventing architecture in response to nature's lack of comfort; although this is, to a certain extent, untrue since humans invent architecture for other reasons and, obviously, nature isn't manifestly uncomfortable). With this free, autonomous nature, we must be respectful but also watchful in view of the dangers that lurk there.

Then there is anthropic or anthropized nature, which is nature that has established a pax rata fiat, an agreement, with the human species, allowing it to inhabit – but not destroy – the natural environment. It's in this context that most of our work takes place: creating links between the two, establishing mechanisms to help nature reclaim its own structures, improving them, restoring them or prolonging them in places where they have disappeared and, in turn, allowing nature to help our species to improve its way of life. It's essential that we be precise when addressing these issues in the projects we carry out.

Martha Thorne [MT] In my opinion, we're at a crucial moment in which the built, natural and digital environments are converging. I think it would be a mistake today to try to separate these ecosystems, given that they're fully integrated. As Joan said, the idea of the built city must be far more closely linked to nature.

During the pandemic we were reminded of the importance not only of natural light but vegetation, open spaces, nature, be it birds, trees, etc. Of how important this is for city dwellers! When I say that we're at a moment of change, it's because I think we're progressing toward the idea that human beings aren't separate from nature but part of the ecosystem we inhabit. I hope we can discover a new relationship with nature, no longer one of management or domination but understanding. The challenge is to understand what our role is from the vantage point of architecture and the built environment, and how the relationship with the natural and digital environment should be, in a more friendly and balanced way.

RD **Martha, you're familiar with Batlleiroig's work. What do you think are the basic features that define its architecture?**

MT I'm really impressed by their background in sustainability and the holistic approach to their projects. It's not about "Let's see how to put vegetation here, then treat the water and then think about how to include users". They seem to me to be pioneers in this field. They talk about the water cycle, about the recovery of nature. When planning everything from bike paths to parks to nature in the city, they always connect not only with the individual but with other phenomena as well, such as water, people, food, sustainability itself. In that sense, I think we can learn a lot of lessons from them. I look forward to seeing what they do in the future.

RD **As you both pointed out, nature is more than what we understand by natural elements. Martha mentioned the importance of the digital, for example, with sensors that allow us to collect data from these realities, and even how with technology we can create new materials, introducing reactive, biological elements. In what ways can architecture be part of this expanded ecosystem? How important is it that a building is also understood as a biophilic artifact? How do you introduce what is natural into office projects?**

JR The quest for biophilic behavior in buildings has to do with the idea of merging architecture and nature. The great garden builders of the past such as André Le Nôtre and Capability Brown differentiated architecture and nature very clearly in their work. It wasn't until post-Romanticism, with Alphand and Olmsted in landscape, or Gertrude Jekyll and Vita Sackville-West in gardening, that environmentally sensitive architecture and nature with biophilic roots began to interact.

When we began to build parks and public spaces in the 1980s, it was customary in Barcelona to follow the tradition of Italian hard squares, i.e., paving the entire surface and relying more on elements linked to urban furniture than arboreal vegetation. Our position at the time was instinctively the opposite. Our first project, the competition for Escorxador Park, was a large park with post-Romantic features, while the second, the Roques Blanques Cemetery, consisted of landscaped agricultural terraces in which the niches, which served as retaining walls, were protected by pergolas. In the nineties, when we began to build offices, we realized that the Anglo-Saxon model of airtight buildings with close control of humidity and temperature wasn't suitable for our climate. This is how we began to introduce terraces, protections and double skins on the most exposed façades, ventilation systems in the windows and vegetation whenever possible. It's this vegetation that allows us to introduce biophilia as a mechanism of biodiversity in the city.

We also introduce biodiversity into parks through amphibian ponds, insect hotels and, in general, with vegetation in which shrubbery abounds. This increases the presence of fauna (butterflies, bees and insects in general), which enhances the environmental life cycle.

RD **Martha, as Dean of the IE School of Architecture and Design, how do you think**

the inclusion of nature in architecture can be taught?

MT I believe that if we want to be on this planet tomorrow, we must protect biodiversity in its entirety. Species not only of flora but also of fauna are being destroyed because of human behavior. Bees are fundamental to the pollination of plants and the health of vegetation, and it has been shown that if one part of the animal kingdom is lost, the danger to humans is direct. It's essential that schools of architecture understand how nature works. We need to study it and respect it to obtain that knowledge. In fact, there's a worldwide trend, perhaps less prevalent in Spain, to change the name of schools: they're no longer called schools of architecture but schools of the environment or the built environment. The programs being developed alongside more traditional careers delve deeper into subjects such as sustainability, natural sciences and engineering, which address topics in the field of agronomy and infrastructure. Regarding Joan's comments about landscape, I think a major evolution is taking place. What we used to call landscape architecture is now seen not so much as a separate discipline but linked to natural flora, fauna and water cycles. The understanding of these cycles in nature merges directly with more traditional studies.

JR Yes, the concept of landscape architecture comes from the Anglo-Saxon world and, particularly, from schools influenced by the teachings of Frederick Law Olmsted. However, Enric and I, while we consider our work to be within the field of landscape architecture, are more influenced by the current French landscape school, by people like Jacques Simon and Michel Corajoud who somehow managed to work not so much with the form of the landscape as with its structural elements: water, land, vegetation and paths. When it comes to landscape, it seems logical to begin by asking, for example, about the behavior of water. What is the rainfall like in the place? Where does the land drain? Where will the water come from to irrigate? The challenge is to transfer this attitude to planning and architecture. That's why we try to design and develop buildings based on their structural elements, hydraulics, vegetation, materiality and routes. Nowadays, it's inconceivable to design a neighborhood or a building without thinking about its hydraulic, climate or biophilic behavior. In fact, this is the only way to respect the structures of the place and, therefore, relate empathetically to the environment.

RD **It's interesting that you apply these commitments equally to these three scales: to territory or landscape, to the unit of public space or urban environment, and to buildings. I believe the most innovative vision is that the scale is indifferent to the application of these systems or design strategies, that in the end what you want to do is to provide solutions to the climate emergency. Given the new post-pandemic circumstances, how do you approach the issues of health and wellbeing at Batlleiroig? How do you design by putting the user at the center, and how does that change the way of designing and understanding the natural elements and all the non-human entities that form part of that ecosystem?**

JR I believe the goal is to put nature at the center. When individuals are at the center (as Le Corbusier exemplified with the Modulor), everything revolves around their comfort but, unfortunately, at the expense of the balance of the surrounding natural environment. When nature is at the center of the equation, respect for it ends up benefiting humans. By focusing on the environment, we're protecting those who live in it. This change of focus is the greatest challenge we must address.

RD **Martha, how do we include the human factor in architectural studies? How do we train architecture students to think about our well-being and health?**

MT The idea of architecture as a closed discipline that contained the humanities, art and science but had a boundary that needed to be protected is no longer appropriate. I believe the way forward is to understand, dialogue with and gather from other disciplines all that they can contribute to our physical environment. If we're talking about human beings, I think of neuroscience, medicine, biology. By this I don't mean that a career in architecture must be so generalized that we need to understand everything. Yet it's essential to understand how the physical environment works, its holistic relationship with the natural and digital environment. Today, a person or a studio that thinks knowledge of other sciences or other disciplines isn't necessary won't be able to take full advantage of the enormous potential of architecture. In fact, if we extend the word "sustainability" to "climate change", we see that it's not about sustaining what we have: the important thing is to understand how to face a future in which we're really in danger and how we can mitigate the damage that has been done. I don't mean we need to take a moralistic or ethical stand. I'm not pointing a finger at any culprits. There are many paths to the future, but they must be enriched by other disciplines and by a greater understanding of the impact that architecture, buildings and infrastructure have.

RD **But in our discipline, we want to create something beautiful as well as sustainable and more user-friendly. Therefore, we must redefine some aesthetic patterns. If, for architecture, the focus should be on integration with nature, what aesthetic patterns can emerge from that? For example, some people demand a beauty based on the laws of thermodynamics.**

JR It's clear that, given our age, Enric and I belong to a generation educated according to aesthetic criteria indebted to both modernity and post-modernity. But in response to the way society has evolved, we've been able to react and put environmental values before compositional ones in our work. However, unlike architects who have made this attitude their rallying cry, we prefer to combine efficiency and sustainability with a respectful image in harmony with the context. We aren't so much interested in obeying these "laws of thermodynamics" as you define them as in adapting our projects to the context, in terms of scale, texture and the history of the place. It's hard not to admire Marinetti's enthusiasm about the aesthetics of the railroad or the airplane or Arseny Avraamov's symphonic compositions with factory sirens, but we're far removed from the aesthetics of revolution. We feel closer to a serene beauty linked to the correct choice of materials, to the logic of its construction and to the efficiency of its environmental performance. The ethics that emerge from constructive logic and judicious interaction with the environment are our most solid working parameters.

MT I like what you said about beauty not being a purpose sought out from the beginning. However, architecture must be about more than efficiency, functionality, or the response to a client's wishes, more than a service profession. It must have the human side that distinguishes it from what technology or artificial intelligence can do. It must incorporate a message that is able to resonate with the community and society, both in the short and long term. In conversations with other school directors, I always tell them: training students to win a Pritzker Prize would be the worst thing to do. If it happens, it's always positive. But to make it an objective from the beginning is a bad idea because it's a value that's so ephemeral it lacks the resistance a building needs, not only in itself but in people's memory of it. I totally agree with the idea that beauty should not be the starting point, but also that the expression of a beautiful form, sometimes personal, sometimes communal, needs to contain a human message, something in its expression that allows it to communicate and resonate with people.

RD **Martha, if you'll allow me to comment on your past connection with the Pritzker Architecture Prize. The fact that the last award was given to Lacaton & Vassal is indicative of a change. It was a clear commitment to an architecture that proposes this shift toward the awareness of a given context, which you take on board, but, at the same time, you want to improve, focusing on the person and the environment. In that sense, Batlleiroig is committed to recycling and how important it is to renew, recycle and restore. In fact, several of your projects have been based on**

existing buildings. It has been proven that the construction of a building consumes more energy than its subsequent use in years. How much do you think it helps and what kind of impact do these projects have on Batlleiroig? How many projects are new buildings and how many are based on opportunities to recover existing constructions, whether they are buildings, landscapes, parks or urban environments?

Doctors Without Borders headquarters in Barcelona, 2017-2019

JR In building construction, it's clearly much more interesting, both from an economic and environmental sustainability point of view, to refurbish existing buildings than to demolish them and build new ones. This is the approach our studio takes. An example is the new headquarters of Doctors without Borders in Barcelona. It's a modest but efficient project, which is why I think it's one of our best. Especially because it doesn't pretend to be anything other than what it is. Often the problem with architects is that they want to make every work their great work. It's a common mistake. They're missing the point of their assignment: they're working for themselves instead of working to solve the problem in all its aspects.

In landscape projects it's more complex. When the place has been practically destroyed, we must restructure it completely, as we did, for example, in the Vall d'en Joan controlled landfill site, in Garraf Natural Park. If the landscape is being destroyed, but there's still a chance to save it, we have to focus our energies on its restoration. In this case, we need to be decisive and recover the original structure. That's always better than creating a new landscape and putting up a new building, because when you recover what's there, you reclaim its values and prolong its life.

MT Unlike other years of the Pritzker Prize, Lacaton & Vassal embrace the very clear and powerful message of "less is more", and that of observing the context with restraint and care. Also, the jury members highlighted the winners' democratization of architecture: a kind of awareness in which architects must understand and improve, as much as possible, what already exists, to give new life to what is already there.

Continuing with the subject of recycling existing buildings, the concept Joan mentioned of giving them a new life is very interesting. When I look to the future and the students in my school, who are now in their twenties, I hope that in ten years, when they're working, there won't be any difference between waste, between the old and the new. I want them to understand they can recycle the windows and many of the materials we use, that the key is not to use materials that can no longer be recycled. I want them to see our buildings as a kind of repository of resources for the future. In that way, using an existing building and everything inside it for a future project is taking sustainability beyond the position of merely protecting what we have. It means rethinking the materials in a building, the use we make of them, and recycling for a purpose. This is where architectural creativity enters the equation. Despite Louis Kahn's statement that a brick wants to be a brick, I think sometimes it wants to be something else. I think this represents a great challenge and opportunity for architects looking to the future.

RD **A characteristic and requirement that we demand of buildings is that they consume fewer and fewer resources and, if possible, generate their own energy. However, to what extent can a building be truly self-sustainable? Can it generate enough energy for itself and for the neighboring buildings? We need to be honest when explaining that, due to**

climate conditions or needs, it isn't always possible to obtain enough supply from renewable energies. The same is true for food demands; putting a vegetable garden on a roof doesn't cover actual consumption needs. What is your critical opinion of these myths regarding energy self-consumption and self-productivity?

JR When you design a building, you must ensure it has maximum sustainability, that it produces minimum emissions, that it has a suitable ventilation system, a high degree of passivity, and good orientation together with high solar protection. You need to use materials that reduce its heat footprint, ones that aren't refractory but absorb that energy. Still, you'll probably fall short of the zero-emissions target. To achieve real self-sufficiency, you should work with groups of buildings functioning like trees in a forest. Just as a forest – a group of trees – works better than an individual tree, buildings work better collectively than individually. By becoming a collective energy production system, a neighborhood can achieve some of the improvement goals that have been talked about so much since the pandemic: accessibility, the 15-minute city and the proximity of resources. When they asked us to isolate ourselves at home, they should have asked us to isolate ourselves in our neighborhood. The neighborhood would have had its own sufficient sustainability structure. If, through agreements between the public and private sectors, we can transform existing neighborhoods, implementing collective energy production systems, increasing the vegetation system to improve air quality, altering the heat footprint and providing them with their own services, we will have more sustainable urban structures. These are the objectives we're currently working towards in the ecodistrict at the former Mercedes factory in the Bon Pastor neighborhood in Barcelona.

LaMercedes ecodistrict in Barcelona, 2019-2027

MT I admire that point of view very much. The context isn't ideal yet, but each building can do its part. I think that based on the scale Joan is talking about, we may well return to the scale of the individual. It's one thing to improve the building, the notion of the adjusted machine, but we also need to think about how we use buildings. I always try to explain this concept to students by comparing it to a car: you're alone in your car for half an hour, park it for eight hours and take up space, then drive it another half hour home where you park it overnight. What a waste of resources! It's absurd to take up everyone's space for such minimal individual benefit! Regarding buildings, a set of political and cultural measures should be taken. We must look at how individuals behave and, based on that, think about their behavior in a new way. Why does a school function for eight hours only? Why can't it be something else afterwards? We saw during the pandemic how our homes had to become offices, schools, meeting rooms, etc. Why not extend it further and make better use of what we have? In addition, elements in the built environment demand our attention. An example is data centers, which are brutal heat-creating machines. We need to look at where they are built, why they are built, whether they have a single use as data centers or whether we can combine other uses to take advantage of the heat they emit. This is a series of measures where the first step, no doubt, is Batlleroig's responsible stance of looking carefully at that building with professionalism and a sense of responsibility. From the academic side in collaboration with our profession and with industry, we need to look beyond individual policy measures and take more of a guerrilla whistle-blower approach, where a problem can be an opportunity.

RD **It's true we can build in a more efficient, sustainable way. But if our habits continue to be to travel alone by car, not to recycle, not to consume zero-kilometer products, etc., we obviously won't solve our problems simply by building sustainably. For example, today the sustainable mobility model being proposed in the 15-minute city, or the idea of the super-block being implemented in Barcelona, doesn't affect people's habits. I'm interested in knowing how to influence habits at the scale of the building, the street and the public space. How can we implement sustainable mobility in the projects themselves?**

JR It's clear that we must change in many aspects, and the main one is probably our notion of comfort and how this is linked to a type of mobility, at the scale of the city and the building. In the urban environment, the replacement of the automobile by collective systems or by non-polluting individual systems (bicycle, skateboard or scooter) is part of this change. In this sense, small-scale operations such as the pedestrian and bicycle connections we made between Barcelona and Esplugues de Llobregat, or between Barcelona and Montcada i Reixac, exemplify how certain changes in road systems can improve urban mobility.

Pedestrian connection between Barcelona and Montcada i Reixac, 2019-2020

Similarly, a given idea of comfort and mobility in a building and exit regulations has prioritized elevators over stairs. This prioritization of the elevator relegates stairs to a secondary role linked only to evacuation and not to the relationship between users. In most of the buildings we design, we try to give vertical routes a social role in keeping with a new idea of mobility and health. This is what we call healthy stairs: stairs that are present, that are visible, and that visually and physically connect all floors.

These small changes don't affect users' comfort but have a positive impact on their health and social interaction.

Building for Zara.com at the Inditex headquarters in Arteixo, A Coruña, 2015-2024

MT I agree completely. Change is needed, but it doesn't have to entail a loss. It can mean giving up something for the benefit of something better that can be an advantage. The pandemic has been dreadful in terms of health and confinement. Yet who would have imagined that our streets would have less noise, fewer cars, more public space, and more leisure. That the city would be used in a different way, for exercise, for classes, as schools, for eating... It's a pity it took a calamity to show us alternatives. I think this example can help us learn how to look at what we have, what we have done in terms of rivers or public space and parks. For example, in Madrid, an abandoned river was transformed to the great benefit of the city and its inhabitants. I look at it with a great deal of optimism. Mobility has many components. In the old days, we thought of it as getting from point A to point B. I believe mobility is, on the one hand, the freedom to use public space, to be able to move around, and, on the other, to do so safely. This doesn't simply imply bike safety or the safety of other vehicles. It means thinking, for example, of our aging European populations, where the user feels less control over the physical space. Sustainable mobility can provide that level not only of comfort but also of safety, of being able to use the city at any stage of life. For me that's what sustainable mobility is: not just getting to your destination but more intangible values such as freedom, belonging, safety and the ability to use common spaces.

RD **Earlier I asked you to define your concept of nature. Now, to conclude, what is your definition of a city?**

MT I believe the city should aspire to be a circular economy; it isn't isolated but connected to the territory, though in a different relationship. Not collecting, extracting, using and throwing away but being in balance with both parts. What I learned from the pandemic is the great opportunity to embrace the city as an inclusive place, where everyone can find their physical and mental space.

JR I agree with Martha. I would argue for a circular, inclusive city but with a certain autonomy for smaller scale sectors: a city of financially sound, well-provided-for neighborhoods that avoid the marginalization of the current peripheries. Obviously, this city must be in tune with the natural structure in which it is located and try to merge with its surroundings. In Sant Cugat, where we worked for years, we managed to preserve the streams as urban parks and maintain their presence in the heart of the city as green fingers. This improved air quality and reduced the heat footprint. It's important for planning to respect the natural structures of the territory in which our cities are located, because, in the end, these structures are the backbone of the city. Nature is the city's soul.

Stradivarius headquarters in Cerdanyola del Vallès, Barcelona, 2016-2017

30 Commitments to Combat Climate Change

Cool Island

Biodiversity

Water

Sustainable Mobility

Productivity

Recycling

Green Life Cycle

Near-Zero Emissions

Self-Sufficiency

Health and Well-Being

10 Concepts

3 Areas

City and Territory

-3°C / Fresh Air Arteries
Parc Central and Open Spaces in Coll Favà, Sant Cugat del Vallès

Green Infrastructure
Llobregat Blue Routes, Barcelona

Original Geography
Riera de Sant Climent Linear Park, Viladecans

Green Routes
Cyclable Connection Between Barcelona and Esplugues de Llobregat Along Avinguda Diagonal

Food Governance
Development of the Finestrelles Sector, Esplugues de Llobregat

Urban Recycling
Provisional Urbanization of the Industrial Area of Can Batlló, Barcelona

City of Cities
Development of the Gallina Blanca Block, Sant Joan Despí

A Naturally Regulated Environment
Llevant Mar Sector, Gavà

Generating in the Neighborhood, Combining Energies
Transformation of Blocks in Barcelona's 22@ District

An Inclusive City
LaMercedes ecodistrict, Barcelona

Landscape and Public Space

Bioclimatic Spaces
Park on the Undergrounded Railroad, Vilafranca del Penedès

New Natures
Landscape Restoration of the Garraf Controlled Landfill Site, Garraf

SUDS - Sustainable Urban Drainage Systems
MEETT Exhibition Park, Toulouse

Door to Door
Pedestrian Connection Between Barcelona and Montcada i Reixac

Zero Km
Boca de la Mina Promenade, Aromas Park and Agricultural Garden, Reus

Revaluing Place
Plaça del Cobert de la Màquina de Batre, Sant Fruitós de Bages

Nature-Based Solutions
Forest Path in Roques Blanques Cemetery, El Papiol

Logical Sustainability
Igualada Central Park

Producing Energy in Public Space
Green Decking of the Ronda de Dalt Beltway, Barcelona

Healthy Public Spaces
Scenic Route in the Old Gypsum Mines, Igualada

Building

Climate Shelters
Ó Building Offices, Barcelona

Biophilic Buildings
Parc Glòries Office Building, Barcelona

100% Reuse
The Style Outlets, Viladecans

Healthy Stairs
Johan Cruyff Stadium, Sant Joan Despí

Productive Buildings
Expansion of Roche Diagnostics Headquarters, Sant Cugat del Vallès

Rehabilitating *vs.* Demolishing
Doctors Without Borders Headquarters, Barcelona

Carbon Neutral Balance
Vega Instrumentos Head Office, Sant Cugat del Vallès

Near Zero-Energy Buildings
Tànger 36 Office Building, Barcelona

100% Positive Energy Balance
Inditex Headquarters, Arteixo

WELL Buildings, the User at the Center
Generalitat de Catalunya Administrative District, Barcelona

Designing the Bioclimatic City

Cities – public spaces and buildings – must be modeled to have all possible connectivities, those that promote ecological, urban and metropolitan connectivity, as well as those that reinforce the social nature of all actions. Only in this way will cities acquire the bioclimatic quality needed to mitigate the effects of climate change.

The different places that make up the city – spaces for living, working, studying, playing or walking – should be interconnected with a series of productive and equipped spaces that promote biodiversity, sustainable energy production, water control, local food cultivation and leisure activities.

This series of projects aims to meet one of the most important strategic objectives for the future of our cities: for pedestrians, cyclists, public transport, food, water, air and, ultimately, life, to flow freely through cooler, more pleasant urban environments.

Cool Island

-3°C / Fresh Air Arteries **Bioclimatic Spaces** **Climate Shelters**

-3°C / Fresh Air Arteries

Green infrastructure plays a key role in the resilience of cities to the urban heat island effect. Open spaces, as major structuring elements of the city, can become bioclimatic spaces capable of regulating temperature and humidity in the urban environment. The linking of these green infrastructures with the original geography, their grouping in areas with sufficient critical mass and the treatment of their vegetation are important aspects that can reduce the temperature by up to 3°C compared to similar environments whose design is based on other criteria.

The study of the surroundings of **the junction of the C-31 Freeway with the C-32 in Montgat** addresses the recovery of the Font Stream as a green infrastructure that supports soft mobility and structures the appearance of the urban fabric with new uses and intermodal public transport hubs.

Preserving the original geography and structuring the city around it

A system of open spaces with a coherent structure, such as natural watercourses, can help to reduce the average temperature in large urban land areas and provide a climate shelter in times of global warming.

Promoting the concentration of green spaces while avoiding fragmentation

The continuity of open space systems increases the cooling effect and helps them to become corridors of biodiversity, especially if they are physically connected to natural spaces. This continuity must also allow for a wide perimeter of contact with the urban fabric that gives it permeability and brings green infrastructure and its benefits closer to the maximum number of citizens.

Maximizing vegetation and choosing absorbent materials

Guidelines for the management of open space systems involve preservation of the natural soil with native species that prevent erosion and create a space for biodiversity, the use of permeable, heat-absorbing materials in transformation areas, and the maintenance and replanting of trees to increase greenery and generate shade. This creates the necessary environmental conditions that encourage citizens to use public space and allow these spaces to extend to the surrounding urban fabric.

Parc Central and Open Spaces in Coll Favà, Sant Cugat del Vallès, Barcelona (Spain, 1989, 1997): an example of how the preservation of the original geography has an impact on the cool island effect.

-3ºC average temperature and **-5ºC** user-perceived temperature in Parc Central compared to Coll Favà

Landscape can be the solution
The Sant Cugat del Vallès park system is the result of the intentional addition of open spaces from a series of partial plans developed at various times with totally different urban planning criteria.
Through the discussion of the different urbanization projects and the establishment of homogeneous criteria, the particularities of each commission have been overcome to obtain a vision that, solely with vegetation and paving, becomes something that seems to predate the construction of the city.

We had the opportunity to participate in the adaptation of two of the sectors with very similar preexisting characteristics to the same urban parameters and green areas.
In the Parc Central sector, we were able to influence the urban planning decisions, and a central green area was proposed to follow the small stream in the area, leading to the modification of the General Metropolitan Plan, which had proposed a large central avenue.
In the Coll Favà sector, a previous urban form had already been determined that placed the buildings in the center of the sector, fragmenting the green areas into different isolated pieces – streets, avenues, squares and various parks – but we also had the opportunity to develop the project for the urbanization of its open spaces.

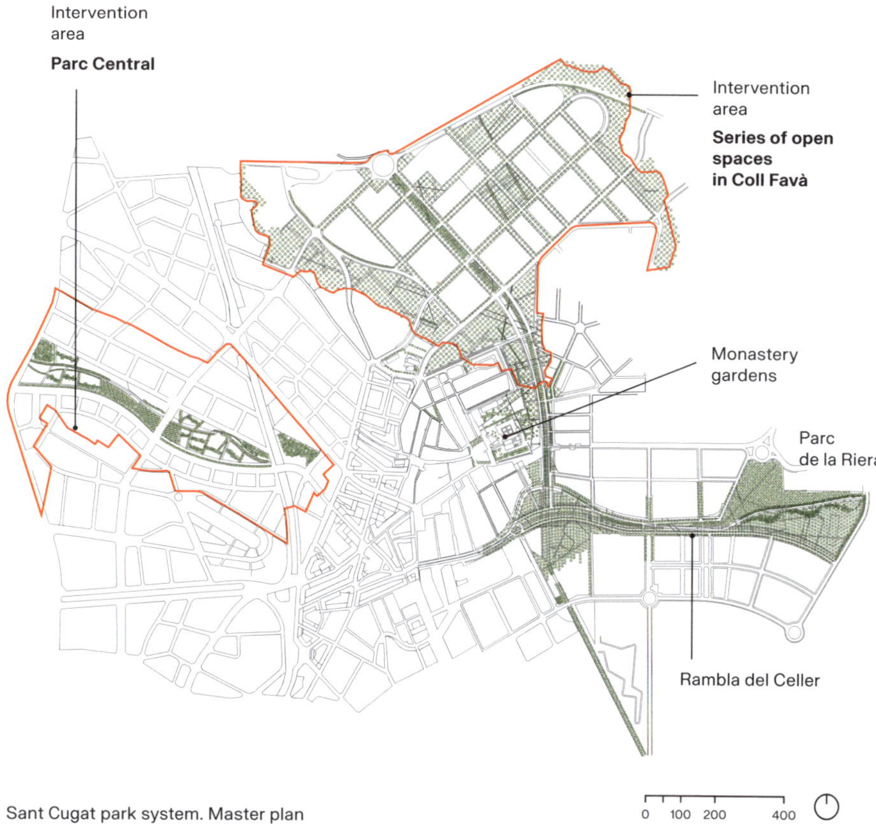

Sant Cugat park system. Master plan

Parc Central sector

Parc Central sector

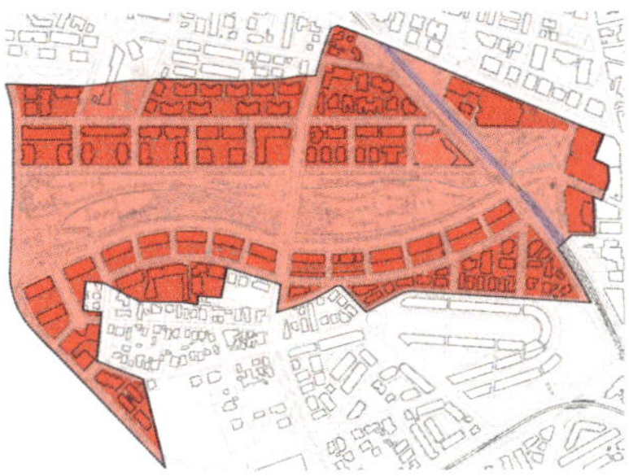

Sector area
32.74 ha

Built-up area index
0.85 m²/m² of land

Open space
53.36%
173,248 m²

37.26% is permeable

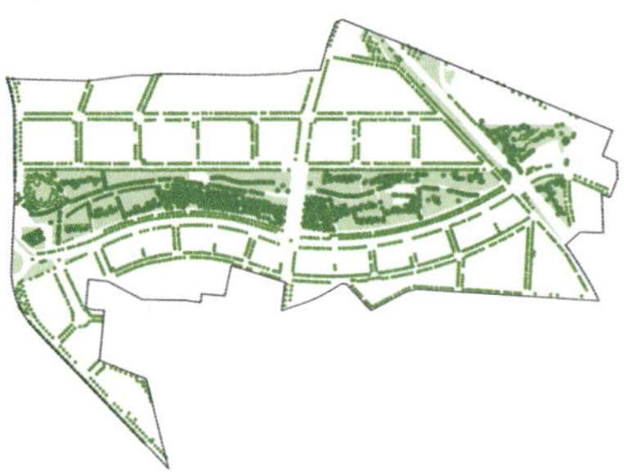

Number of trees / Open space area
1.7/100 m²

Tree canopy area
74,547 m²

Green surface area
64,146 m²

Summer temperature
Day 33.12°C
Night 25.78°C
Average 32.32°C

Coll Favà sector

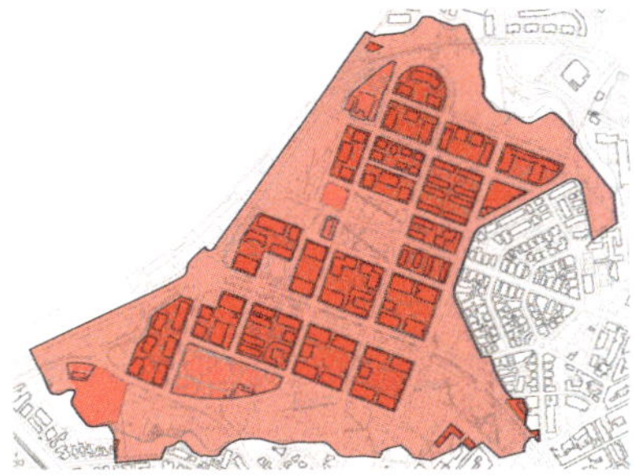

Sector area
56.79 ha

Built-up area index
0.61 m²/m² of land

Open space
59.21%
336,213 m²

53.99% is permeable

Number of trees / Open space area
0.7/100 m²

Tree canopy area
64,878 m²

Green surface area 176,856 m²

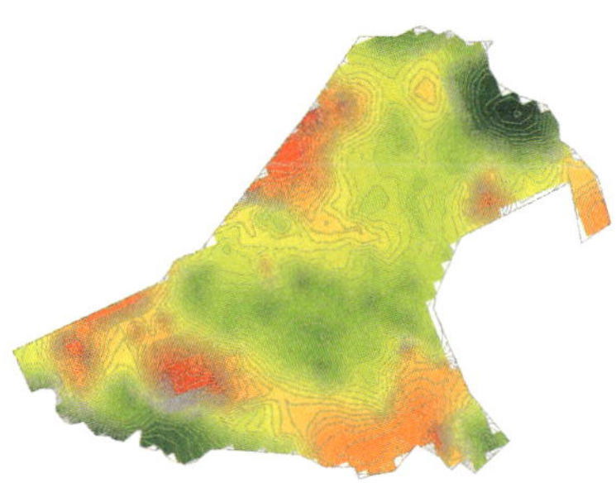

Summer temperature
Day 38.14°C
Night 27.05°C
Average 35.32°C

Urban planning is not harmless
Together with Professor Blanca Arellano and Professor Josep Roca from the School of Architecture of Barcelona (ETSAB), we conducted a comparative study between the two sectors in Sant Cugat to determine the differences between the temperatures in both. The study was published with the title "Green areas and urban heat island: combining remote sensed data with ground observations" in several international scientific journals.
The study shows how two sectors, with similar environmental characteristics and developed with similar urban parameters, but designed based on very different conservation approaches to the original geography, concentration of green areas, selection of trees and choice of paving materials, came to have a temperature difference of more than 3°C. This fact also has repercussions on adjoining built-up areas.

Coll Favà sector

Parc Central sector

In Parc Central the original geography was preserved. The plantations are varied, promote biodiversity and have generated an extensive canopy. The pavements are ceramic on the main routes.

Coll Favà sector

The Coll Favà sector was created new and planted throughout with poplars. The pavements in this area are asphalt.

PARC CENTRAL. Client Sant Cugat del Vallès Council / **Authors** Enric Batlle, Joan Roig / **Team** Lluís Jubert, Lluís Roig, Manel Colominas / **Site area** 32.74 ha // **OPEN SPACES IN COLL FAVÀ. Client** Sant Cugat del Vallès Council / **Authors** Enric Batlle, Joan Roig / **Team** Lluís Jubert, Teresa Galí-Izard / **Site area** 56.79 ha

Bioclimatic Spaces

All public spaces must become bioclimatic spaces able to transcend their own boundaries, infiltrating between buildings and adjacent infrastructure. Designs need to maximize green coverage and shade in public spaces to help reduce temperatures in the sunniest months, thereby establishing bio-synergies with other nearby natural spaces to provide the desired cooling effect in the city.

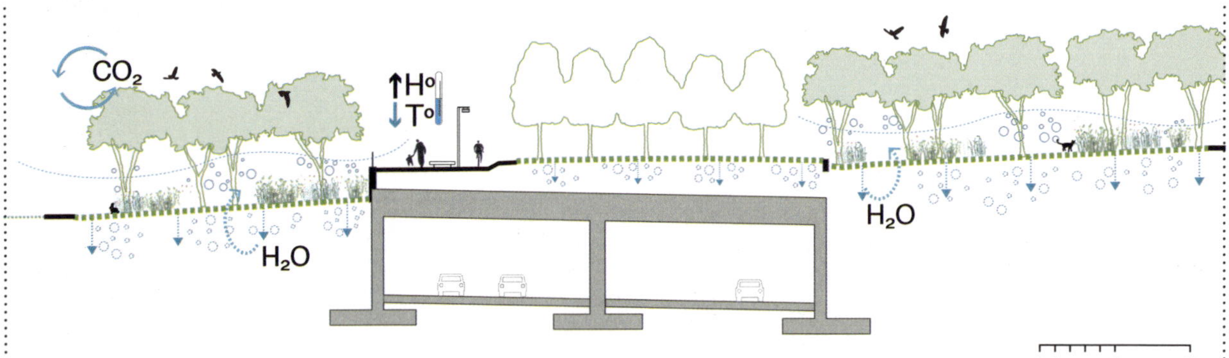

Parc del Tramvia is a good example of a bioclimatic space. Located between the municipalities of Tiana and Montgat in the province of Barcelona, on a major road network (B-20), it involved undergrounding 900 meters of highway with 2 meters of soil and enabled the planting of more than a thousand trees and a 10,000 m² area of large grassy areas for public use.

Thinking of public space as an island of coolness

We have to design our open spaces as pleasant, comfortable, pollution-free places for citizens that improve environmental quality. Designing open spaces to take advantage of the seasonal climate is the key to mitigating temperatures during the hottest and coldest months and facilitating the enjoyment of public space throughout the year.

Naturalizing our cities

We have to prioritize large green spaces and minimize built elements to improve air quality and ventilation, and reduce noise pollution. Trees and landscaped areas must be incorporated into urban designs to regulate the hygrothermal conditions of the environment, relative humidity and the sensation of coolness in the air. Planting deciduous trees in public spaces will maximize sunlight in winter and provide spacious shaded areas in summer.

Promoting the dynamics of natural systems

Distributing the planting of large tree masses will facilitate ventilation corridors in public spaces. By controlling the orography, we can maximize the collection of rainwater and its absorption in green areas. We must minimize hard paved surfaces and establish bio-synergies with nearby natural spaces by means of green corridors to boost biodiversity in the city.

Park on the Undergrounded Railroad in Vilafranca del Penedès, Barcelona (Spain, 2014-2016): a new open space over the railway tracks that introduces the cool island effect through a linear park.

26,000 m² of concrete slab converted into a public park

594 trees planted on top of the undergrounded railway tracks in Vilafranca

60% (15,400 m²) of urbanized area with soft, permeable materials: coarse sand, grass or vegetation

The passage of the high-speed train (AVE) through the center of the city of Vilafranca del Penedès, taking advantage of the pre-existing railway line, ultimately led to the undergrounding of the tracks through the urban center. This involved the recovery of approximately 11 hectares of public space in what had historically been a 1.5-kilometer gap that divided the city in two. The associated technical constraints of working on this railway structure together with significant budgetary restrictions made the assignment a real challenge.

The project is defined as a large linear park in the form of a green corridor that restores connectivity between the two parts of the city and creates a new focal point in the area with numerous uses. The planning proposal maximizes the presence of vegetation and permeable surfaces with the aim of providing Vilafranca with a bioclimatic regulator capable of counteracting the effects of the heat island typical of the city center.

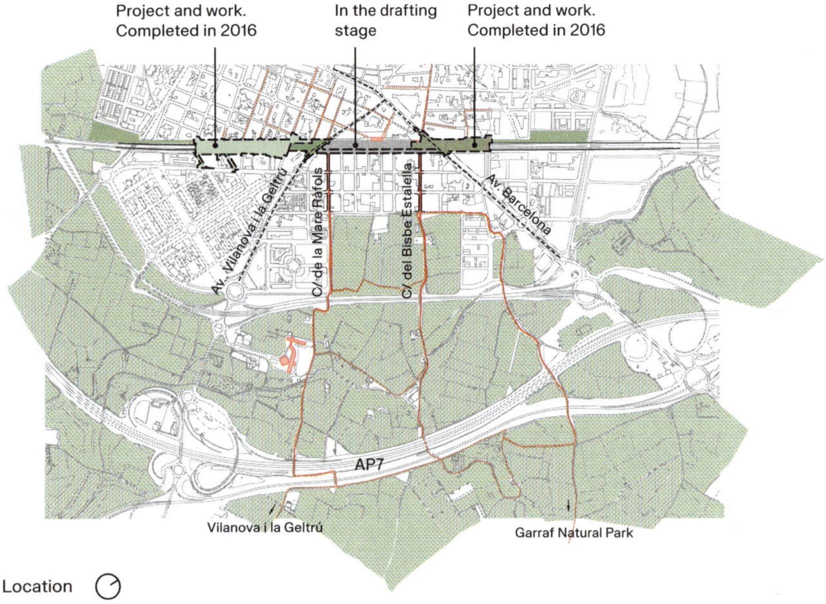

Location

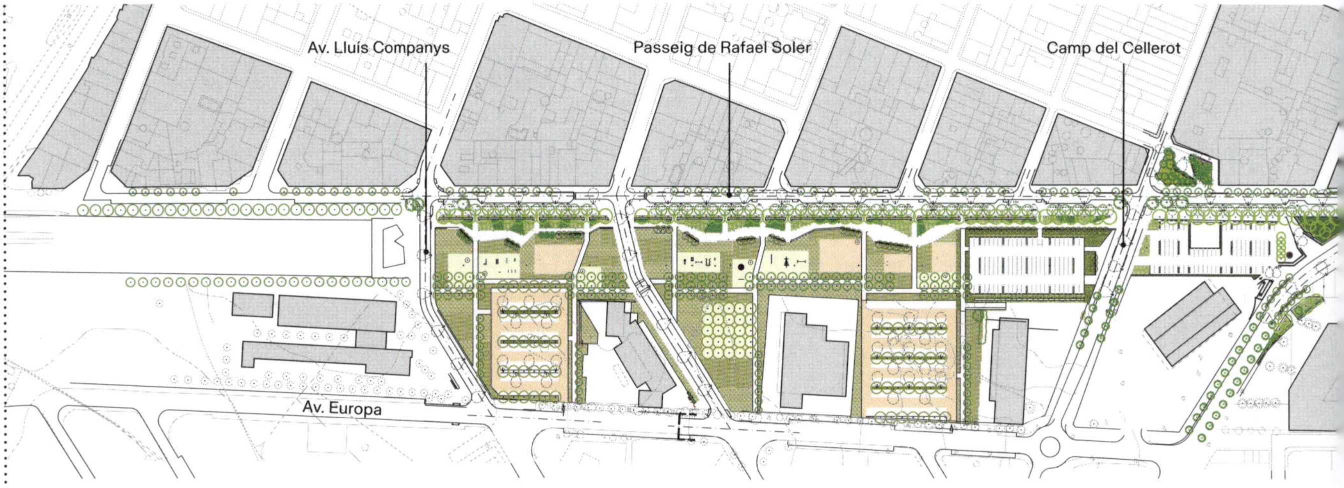

General plan

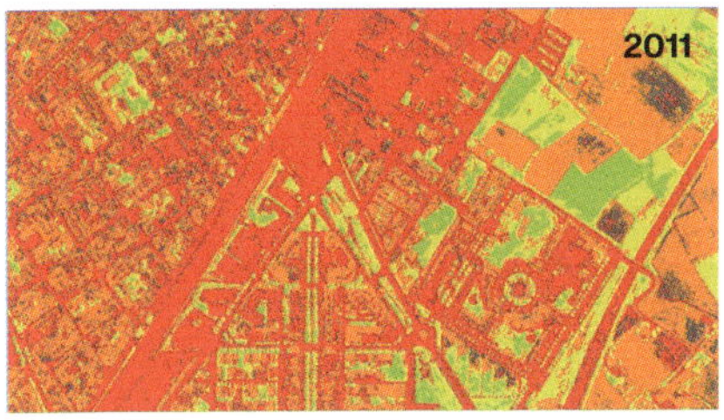

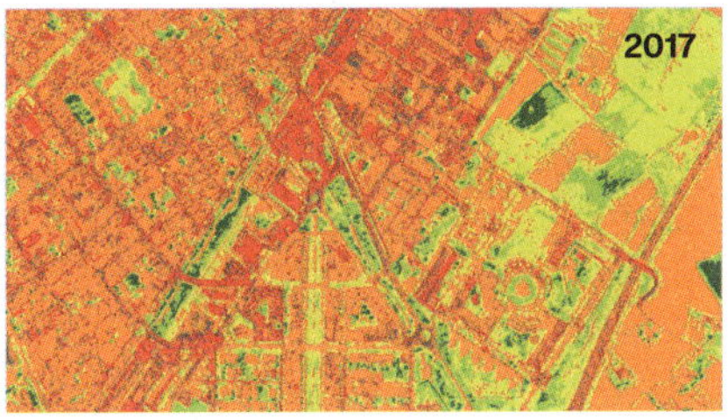

The cool island effect is now visible in the center of Vilafranca del Penedès (2011-2017)

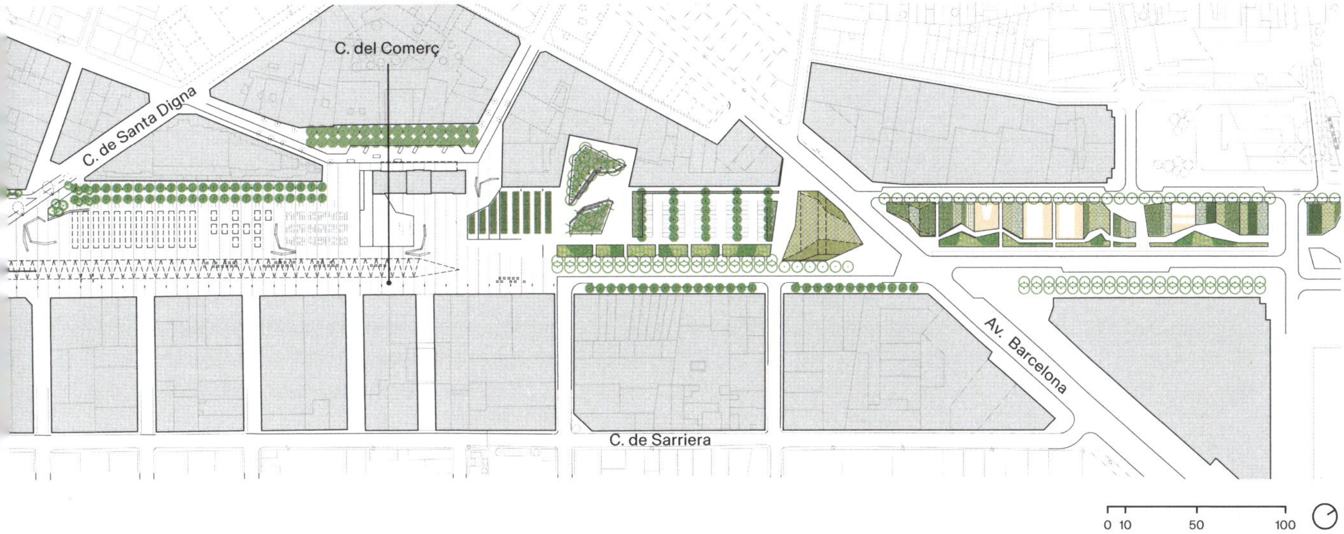

Flowerbeds that absorb rainwater from the roadways

Tree-lined promenade

The side road is transformed into a promenade with more trees and shrub vegetation, integrating the pre-existing pine trees and incorporating sustainable drainage systems.

Garden
The railway covering is transformed into a biodiverse garden where inhabitants can stroll along naturalized paths, surrounded by a variety of shrubs and a huge multipurpose meadow of unique proportions in the city center.

Client Vilafranca del Penedès City Council / **Authors** Enric Batlle, Joan Roig, Iván Sánchez / **Team** Joan Batlle, Livia Valentini, Mercè Lorente, Dolors Feu, Yago Cavaller, Elisabeth Torregrosa, Laura Luque / **Collaborators** SBS Simón i Blanco – Civil engineering / **Contractors** Rogasa, Nouverd, Catalana d'Obres del Penedès, Calaf, Copcisa / **Project area** 39,019 m²

Climate Shelters

Buildings should provide maximum interior comfort, but we also have to design them to have the least possible impact on their immediate surroundings. We can mitigate the heat island effect through shading and the incorporation of cooling elements that, in turn, help to reduce the energy consumption required to achieve indoor comfort.

In **the Àlaba 111 Office Building**, a large porch on the ground floor offers direct connection with the interior of the block, while the office floors include 1,400 m^2 of terraces with vegetation. Both strategies create a new climate shelter for the city of Barcelona.

Improving environmental conditions

Connecting buildings to nearby green spaces and creating as much usable outdoor space with its own architecture as possible and their naturalization – ground floor porches, terraces, roofs, courtyards and gardens – will help to improve environmental conditions.

Proposing correct architectural morphologies

We need to design enclosures with solar orientations that use all types of passive mechanisms – ventilated façades, double skins, vegetation – and are designed with the absorption, inertia and heat characteristics appropriate to the climate in which they are set.

Protecting the building adequately

We can improve the thermal inertia of a building by designing its roofs as useful green spaces where land mass, the increase in insulation and the decrease in reflectance function as a thermal regulator and help to reduce the energy consumption required to achieve indoor comfort.

Ó Building Offices in Barcelona (Spain, 2017-2019): a building integrated into its surroundings that mitigates the heat island effect thanks to its ceramic façade.

50% glazed surface in the façade

7,038 m² of ceramic slats that protect the building and integrate it into its surroundings

946 m² of landscaped terraces that provide coolness and biophilia

LEED GOLD
Certification

The Ó Building office block is on the former Can Batlló factory site, one of the focal points of development in the city of Barcelona, directly related to the economic district of the municipality of L'Hospitalet and important road axes such as Gran Vía de les Corts Catalanes and Avinguda Carles III.
The building contains a central volume in the form of a 12-story tower, where the main offices are located, and a high-ceilinged first floor area, where the different commercial activities take place, freeing up part of the plot's private space for public use.

General floor plan of the Can Batlló site

Historical heritage of the Can Batlló site
The building respects and is integrated into the Can Batlló factory environment thanks to the use of recognizable materials such as ceramic in the façade and in the urbanization of outdoor spaces.

Architecture and city are related in private outdoor spaces for public use. The building creates public areas that connect it to the internal circulation system of the Can Batlló site and to Gran Vía through landscaped spaces and a sheet of water that thermally regulates the entrance areas. The design of roof terraces, the creation of landscaped spaces and the use of reflective paving enhance the cool island effect, as well as providing spaces for users to interact.

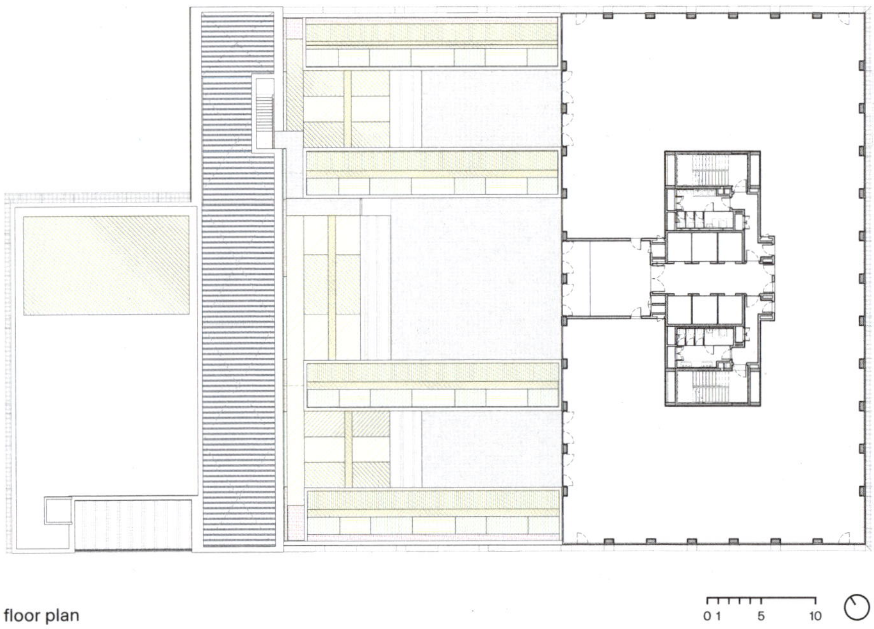

1st floor plan

In the space of a few years, we have gone from constructing buildings in which the structure was moved inward and the ceiling was reduced to achieve 100% glazing on the façade to restoring the structure to the façade and installing solar shading. In the Ó Building, the optimization of glazed surfaces has been encouraged by minimizing the building's energy losses and gains, with direct repercussions on its exterior temperature. Replacing the structure on the façade led to a 50% reduction in glass surfaces while maintaining the transparency desired by users. This solution also achieves a totally open-plan office floor of 1,390 m^2.

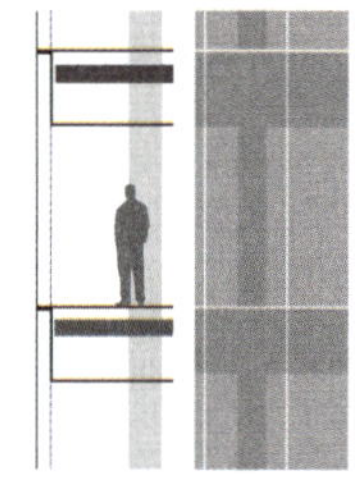

Luxa building, 100% glass

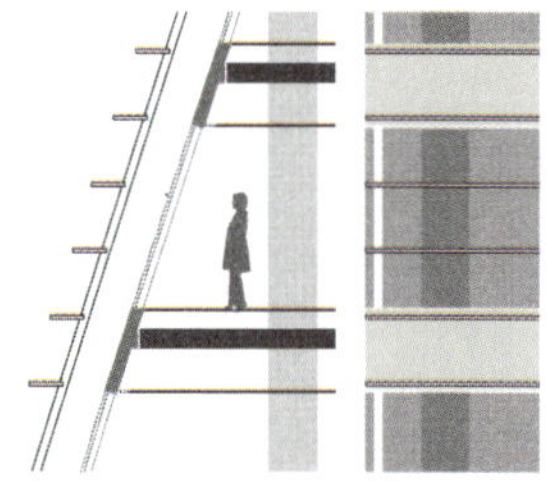

CMT building, 75% glass

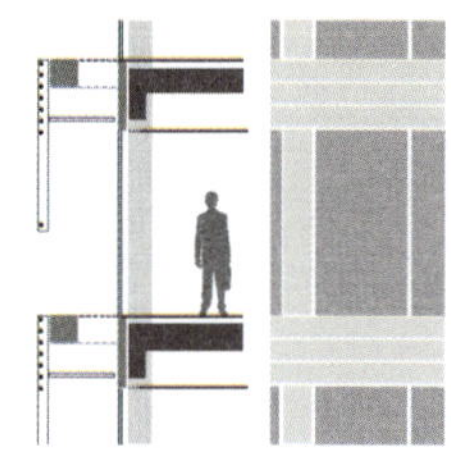

Ó Building, 50% glass

Evolution of the façade glazing solution in different Batlleiroig projects

A sustainable and efficient façade

The double ceramic skin façade enhances the building's energy performance. Thanks to the work of the module and the full-empty relationship, the façade takes into account the different orientations of the building to ensure solar protection, air renewal and visual permeability for users inside.

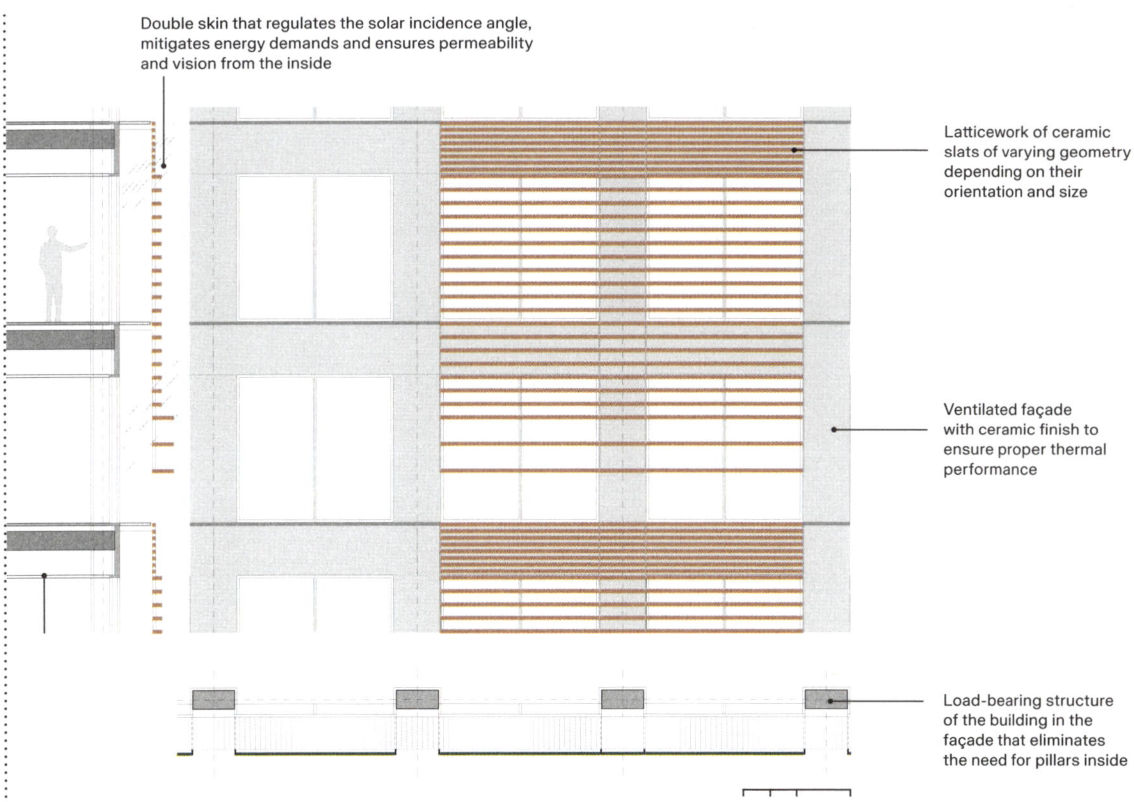

Client Green Clover Capital, BeCorp / **Authors** Enric Batlle, Joan Roig, Albert Gil / **Team** Meritxell Moyà, Antoni Duran, Diana Calicó, Marc Torrella / **Collaborators** STATIC – Structural engineering, PGI Engineering – Facilities engineering, Rafa Capdevila (Bardají Capdevila Management) – Building engineering and construction management / **Contractor** Promuobra Proyectos y Obras / **Gross floor area** 32,776 m²

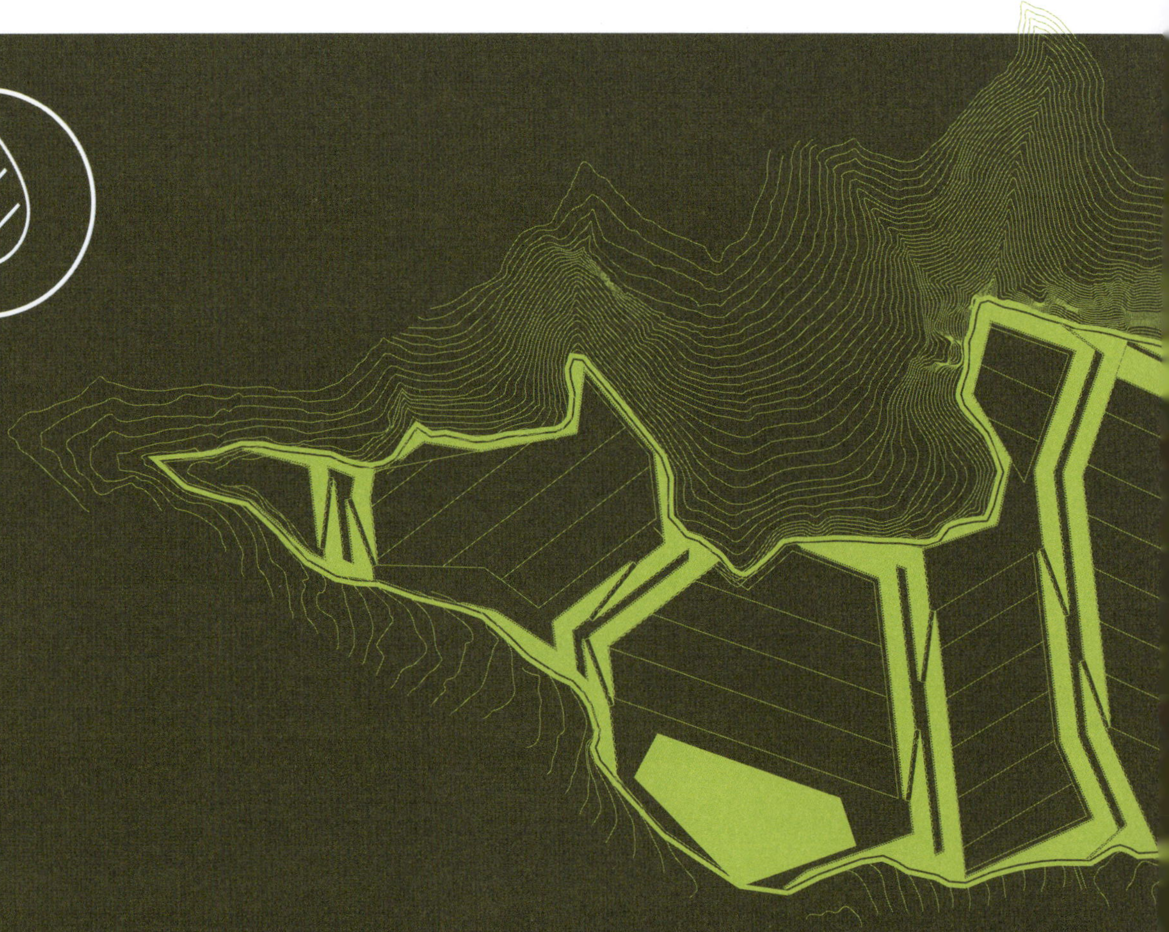

Cultivating Life

Creating new natures is possible. We need to move beyond the discussions of how to conserve landscapes – landscapes that frequently have lost the processes that sustained them. It is possible to recover life in all places, even those where it has been totally lost.

Michel Corajoud defined our profession as follows: “To be a landscape architect is to design landscapes, to prepare places that will create themselves over time or through previously established processes.”

We must avoid the loss of biodiversity. We can promote fertility and establish systems that help us to cultivate life.

Bio-
diversity

Green Infrastructure

New Natures

Biophilic Buildings

Green Infrastructure

Wide-ranging green infrastructures occupy a significant part of our urban landscape, a network of open spaces full of ecological values, possibilities for leisure use and productive capabilities. This environmental and social network could be the backbone of our cities. It consists of a series of spaces with different characteristics but a common objective: to foster urban biodiversity, promote ecological and social connectivity, and enhance their productivity to guarantee their conservation.

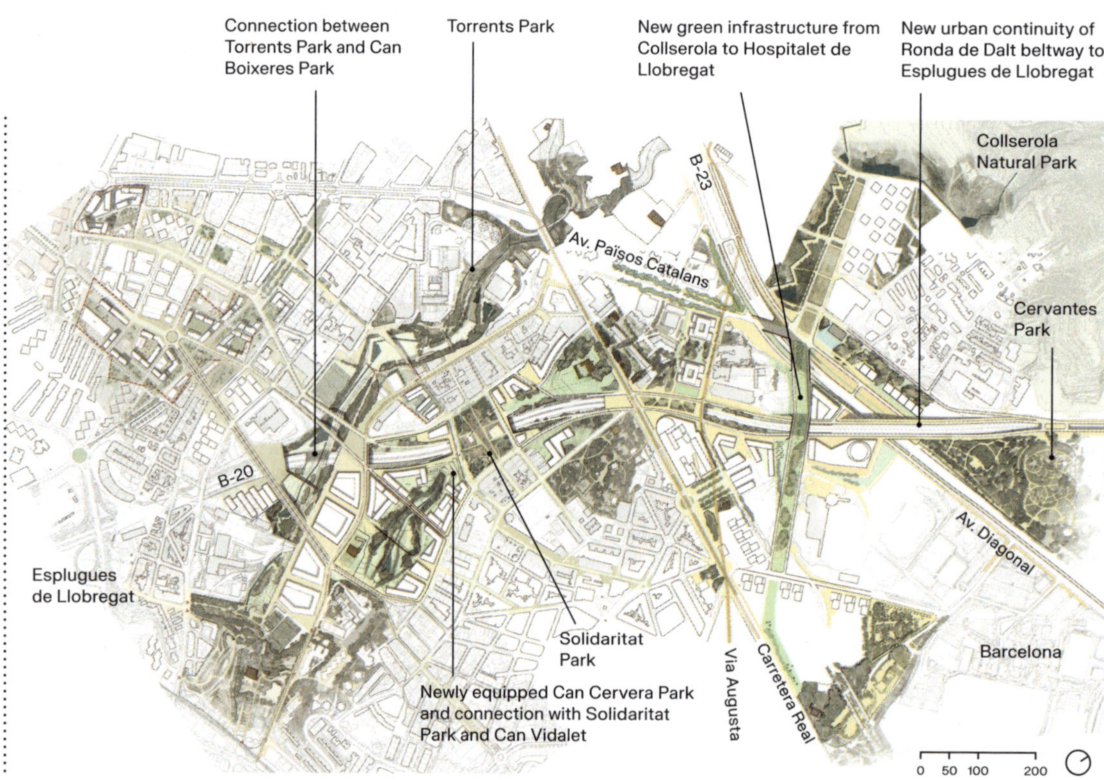

Transformation Studies of the Surroundings of the B-20 Freeway in Esplugues de Llobregat point to improved urban integration based on the maintenance and recovery of natural green corridors, formed in this area by the fast-moving streams flowing down from the Collserola hill range.

Preserving and enhancing metropolitan biodiversity

The ecological matrix that makes up the network of agroforestry mosaics and the hydrologic system must enter the metropolis, and maintain or reinforce existing or recovered biodiversity spaces.

Enhancing the ecological and social connectivity of green infrastructure

The system of green infrastructures in our territory should be not a sum of isolated green elements but a system of networks that deliberately overlap to allow the various ecological, urban and territorial flows to circulate in harmony and function as a single system that enables all kinds of human and ecological connections.

Placing value on its productivity to ensure the conservation of its extension and quality

In addition to its roles as preserver of biodiversity and civic meeting point, green infrastructure should maintain and reinforce the traditional role of a space for agriculture and livestock, bringing food production closer to consumers. This proximity should not only be physical; its objective should be for citizens to get to know and enjoy agricultural areas as a place for recreation and learning.

Llobregat Blue Routes, Barcelona (Spain, 2006-ongoing): a green infrastructure that recovers its intrinsic environmental values and promotes socially responsible use as a peri-urban open space.

270 km of Blue Routes between the source of the Llobregat River and the sea, and along the Anoia and Cardener Rivers

Historically, the Llobregat River was the most popular corridor for new infrastructures that enabled the development of new land. The metropolitan area of Barcelona is perhaps where this phenomenon was the most intense. Over the last few decades, territorial roads, freeways, railway tracks and all kinds of service networks have been built. The result was the overall deterioration of the river as a natural area, its living conditions and its water quality. Furthermore, the continuous infrastructure along its course created an enormous barrier that in many cases severed the connection between the towns and their inhabitants with the river.
In recent years, our attitude as a society toward the river has changed. We want to reclaim and improve its ecological values while, at the same time, taking advantage of its immense potential as a recreational space and contact point with nature in our cities. Within this context, we have had the opportunity to participate in several projects, always with an overarching vision, seeking a balance between these two realities: the river as a living and changing thing, full of life and with its own dynamics, and, simultaneously, the river as a periurban recreational space that the public can enjoy responsibly.

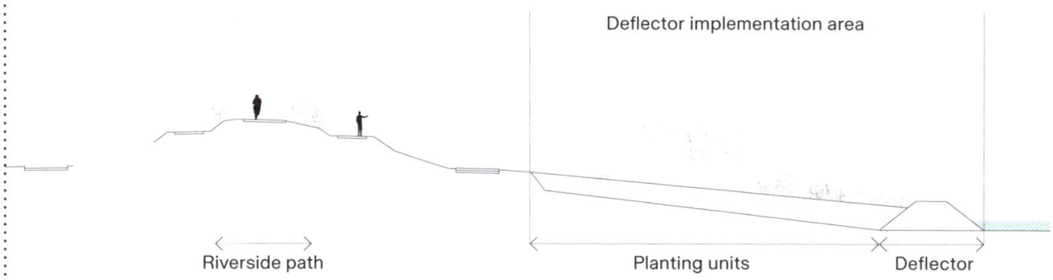

Deflector section detail

Improved environmental value
All the actions seek to improve the environmental values of the river, re-naturalizing its neglected areas, strengthening the presence of river vegetation and even recovering river bends removed or altered by infrastructure.

A network of paths on a metropolitan scale
A new sustainable mobility social infrastructure makes it possible to follow the river along both its banks and connects to the urban centers of neighboring towns.

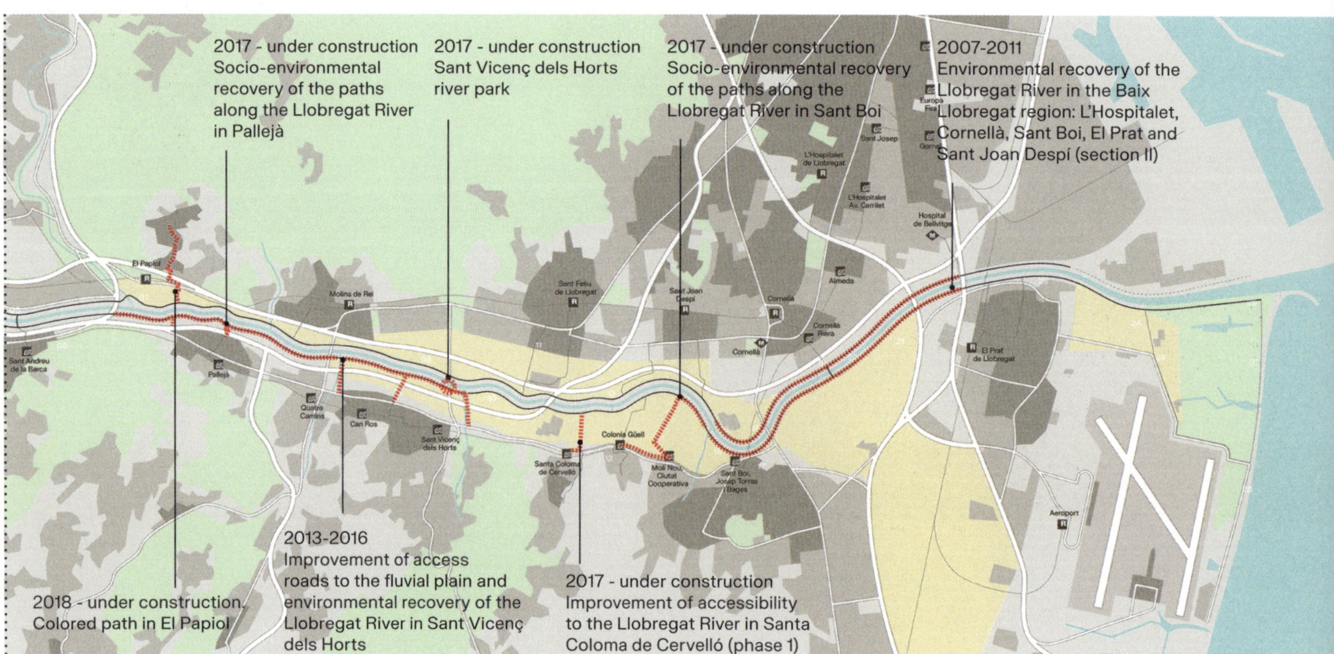

Series of actions carried out on the Llobregat River where it passes through the metropolitan Barcelona

Series of deflectors for recovering bends in the river along the Sant Boi - El Prat de Llobregat section

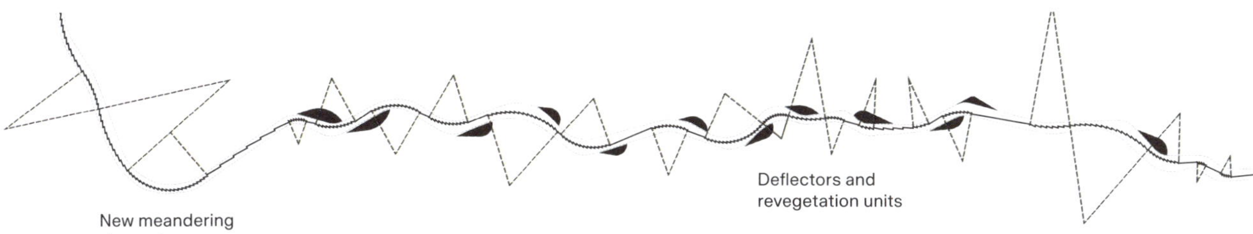

Diagram of new morphodynamics

The Llobregat River in the metropolitan area
A set of guidelines that define the nature of this new Blue Route system (routes for pedestrians and cyclists along the course of the Llobregat River and its tributaries) governs the series of interventions on the Llobregat River in its passage through metropolitan Barcelona:

Strategies for action

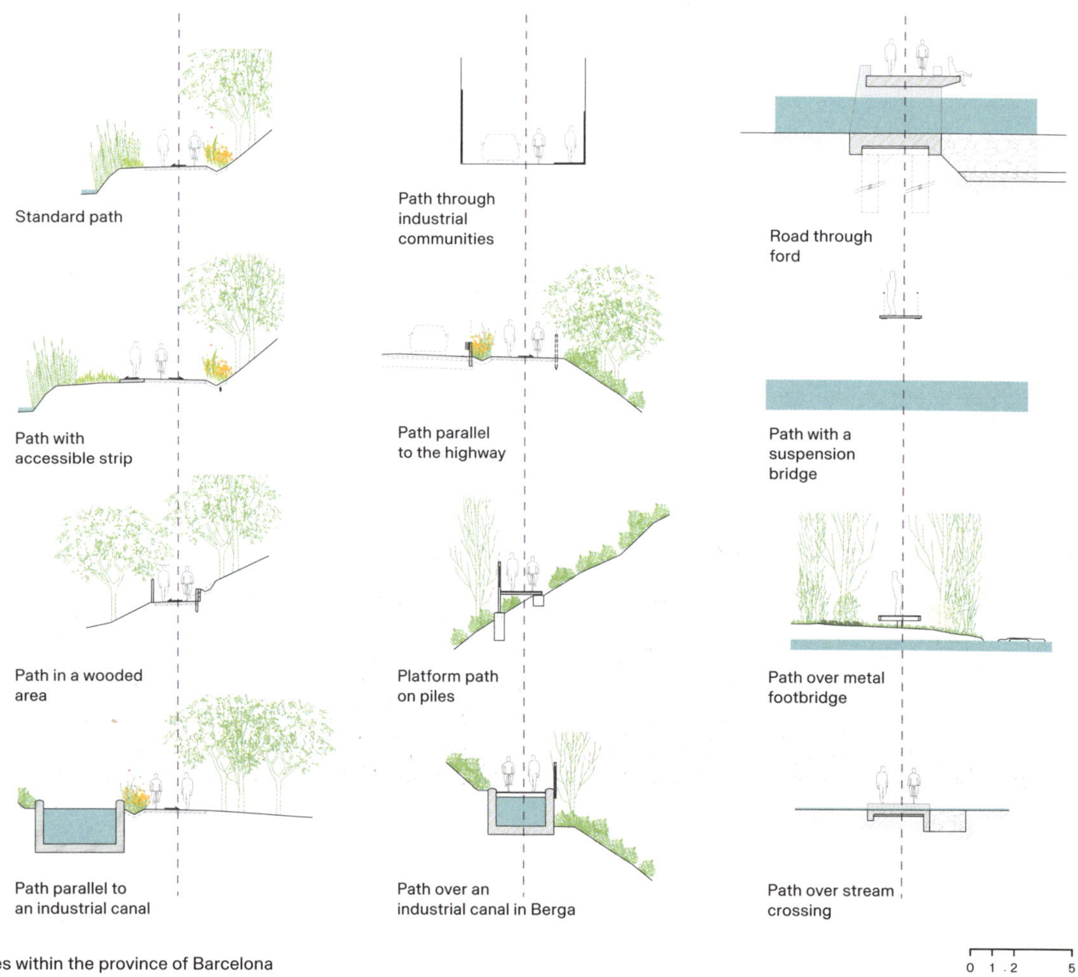

Section types within the province of Barcelona

Gateway to Cervelló stream in Sant Vicenç dels Horts

River gateways

New gateways indicate river accesses. These are meeting points with distinctive language and equipment that clearly identify access to this natural area.

Rest area, Sant Vicenç dels Horts

Rest areas and lookouts

The resulting new network of paths has a series of rest areas and lookouts. The location of these points pays close attention to the values and potentials of the place, including its visual features, pre-existing vegetation and connection points with other paths.

The Llobregat Blue Route as it passes through Montserrat

Since 2017, work has been underway on a project, promoted by Barcelona Regional Council, to make a new leap in scale to this social infrastructure. The Blue Routes project along the Llobregat River seeks to create continuity with the network of pedestrian and cyclist paths throughout the extension of the Llobregat and its two main tributaries, the Cardener and the Anoia. The new network will consist of more than 240 kilometers of paths through five counties and 51 towns in the province of Barcelona.

Starting/finishing point of the Anoia-Pont del Dimoni stretch of the Blue Route

Starting/finishing point of the Llobregat - Castellar de n'Hug stretch of the Blue Route

BARCELONA METROPOLITAN AREA. Client Barcelona Metropolitan Area and Town and City Councils / Authors: Enric Batlle, Joan Roig, Iván Sánchez, Mario Súñer / **Team** Dolors Feu, Yago Cavaller, Xavier Ramoneda, Clara Terradellas, Marta Ines, Gabriel Romero, Enrique Valiente / **Collaborators** Typsa Tecnoma – Engineering of hydraulic and environmental aspects / **Contractors** Acsa, Dragados, Tau Icesa, Moix Serveis i Obres, Pasquina, Ambitec, Benjumea / **Surface area** 2,200 hectares divided into different interventions along the 30 km extension // **PROVINCE OF BARCELONA. Client** Barcelona Provincial Council / **Authors** Enric Batlle, Joan Roig, Iván Sánchez, Mario Súñer with SBS Simón i Blanco / **Team** Dolors Feu, Yago Cavaller, Clara Terradellas, Marta Ines, Gabriel Romero / **Project area** 240 km of future path network

City and Territory

New Natures

The loss of biodiversity and the disconnection of natural habitats amplifies the risks arising from climate crisis. The solution lies in nature, in its care, regeneration and recovery. A place full of life will always be a healthy and resilient place in the face of environmental adversity.

We need to know about and integrate ourselves into the natural cycles and habitats that surround us, to appreciate what is native, what is spontaneous and the variations throughout the seasons and over the years. We need to design with an eye toward generating synergies between the natural and the anthropic, seeking a paradigm shift in the way we relate to our environment by creating new natures.

In **the Sant Joan Despí River Bend Environmental Recovery Project**, field work was key to confirming that a nearby water treatment plant was allowing a constant flow of water to course directly into the river. After intense coordination work between local governments, it was possible to partially capture this water as a bypass so that, prior to its discharge into the main riverbed, it would feed a system of irrigation channels through which the meander was renaturalized and a series of wetlands was created. This new situation is ideal for grazing sheep, which find food and water while simultaneously maintaining and fertilizing the new meadows.

Promoting soil fertility with the aim of creating places that increase their biodiversity

In our cities, 60% of public space is devoted to vehicles. The remaining 40% is shared by flora, fauna and people. New natures improve this situation, decompressing and renaturalizing the public space to make it healthier and capable of harboring life. Any site, even the most altered and degraded, can be rehabilitated and play a key role in promoting the biodiversity of its surroundings.

Recovering and enhancing the natural systems of places and making their values known

It is crucial that we overcome obsolete design logic based on imposed formalisms or on pursuing idealized images of a lost nature. New natures must be an outgrowth of the knowledge, integration and enhancement of the natural systems of the place – forests, meadows, streams, agriculture – and their evolutionary and changing dynamics throughout the year. New natures must develop their didactic and pedagogical potential to transmit their ecosystem values to users, thereby enhancing the feeling of belonging and identifying with the place.

Symbiotically synchronizing our progress with nature

New natures seek symbiotic relationships with the habitats and ecosystems where they are implemented. The time of making proposals that run counter to nature is over. New natures should be thought of as systems with their own evolutionary logic and dynamics, with the solid objective of becoming ideal habitats for flora and fauna in the most self-managed way possible. New natures complement the ecosystems where we implement them, contributing to their preservation and recovering lost values.

Landscape Restoration of the Garraf Controlled Landfill Site, Barcelona (Spain, 2002-2022): a new agricultural landscape based on topology, hydrology and vegetation that facilitates the gradual recovery of life.

75 ha of landfill converted into a natural park

30 hectares in phases 1 and 2, and 45 hectares in phases 3 and 4

151 biogas capture wells for landfill degasification and generating electric power

In phases 3 and 4

4 naturalized ponds

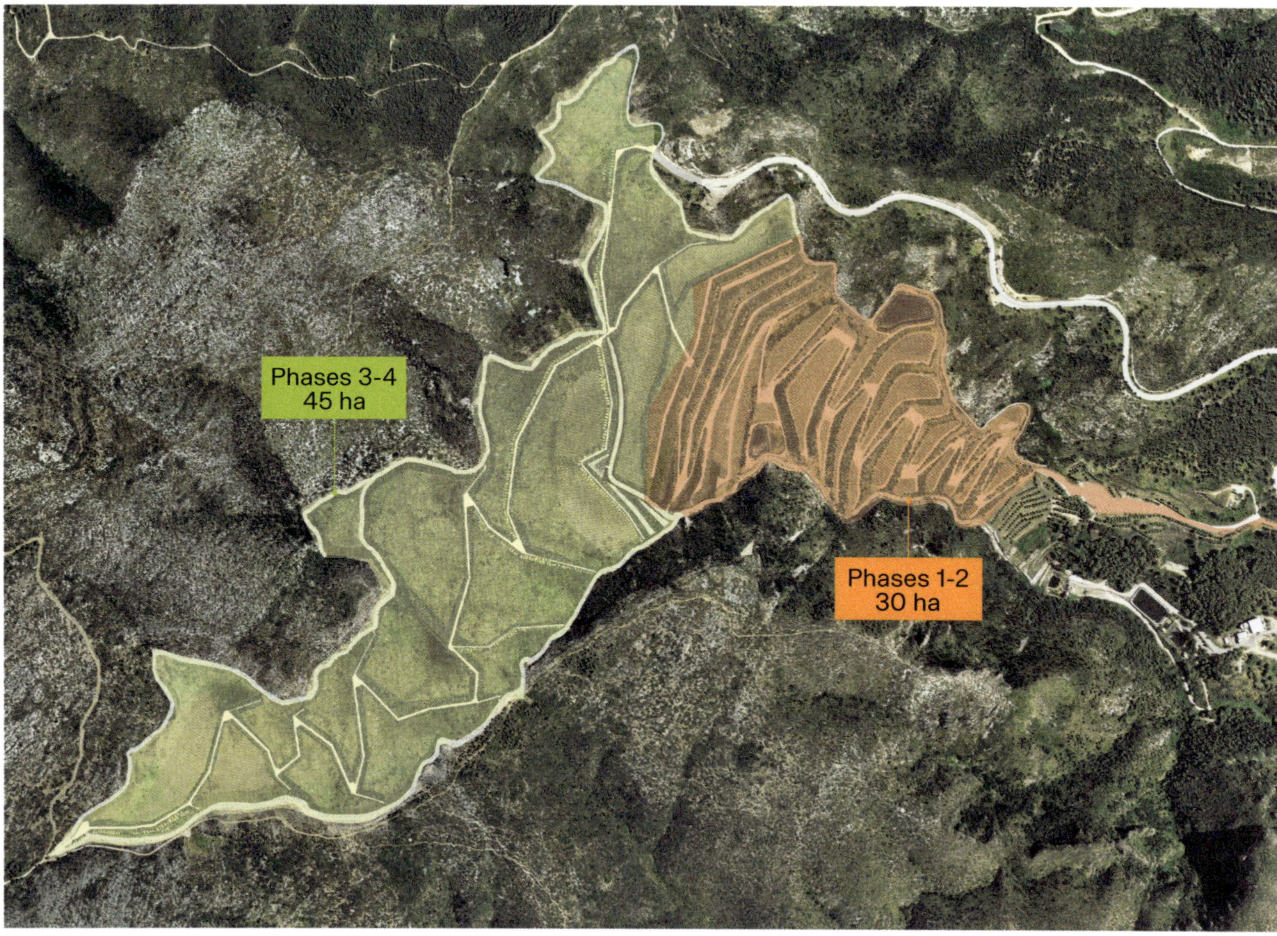

Our landscape restoration project for the old Vall d'en Joan landfill in El Garraf is based on an agricultural system that simultaneously solves the environmental problems created there, the need to create a new landscape and the will to transform it into a new public space.

The decommissioning and recovery in phases 3 and 4 of the Garraf Controlled Landfill Site signal the conclusion of a process that lasted more than 20 years. Yet rather than an end, this represents the beginning of another process, the renaturalization of the area, degasification and leachate treatment so that in the not-so-distant future, it can be an open space that serves the public as the main access to Garraf Natural Park.

Alejandro Aravena, curator of the Venice Architecture Biennale, selected the project to appear in the central pavilion to illustrate how an agricultural project can recover lost biodiversity in a place where Barcelona's waste had been dumped for 40 years. To express our idea of what naturalize means, we chose the motto Farming Life, associating all the wealth inherent in agriculture with the willingness to recover life in a neglected place.

Restoration system evolution

The preexisting topography in the focus area of phases 3 and 4 is gentler and more elongated than the mountainous terrain of phases 1 and 2. This makes it possible to develop a new arrangement where a single ridge runs in zigzag fashion along the length and breadth of the area, meeting all needs. It also makes it possible to structure the more than 46 hectares in a series of 16 meadows, making water management possible through ditches at the foot of the hills that channel the runoff to perimeter drains. The maintenance road that provides access to the entire area through a series of gentle slopes is located on top of the ridge.

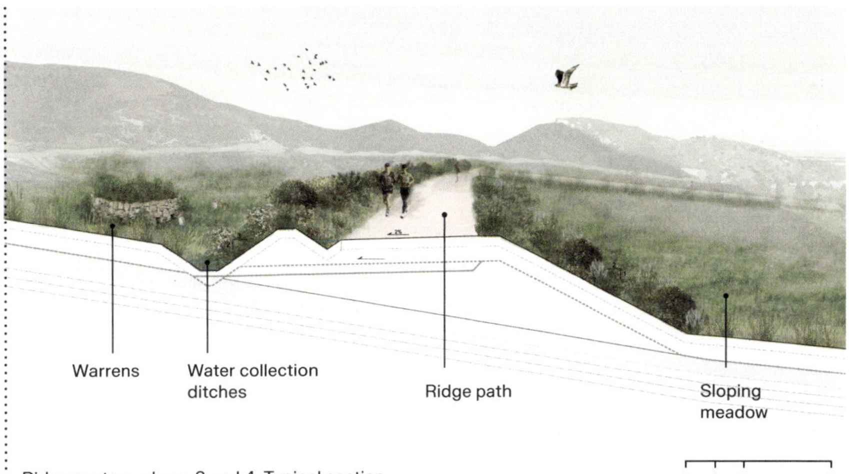

Ridge system, phase 3 and 4. Typical section

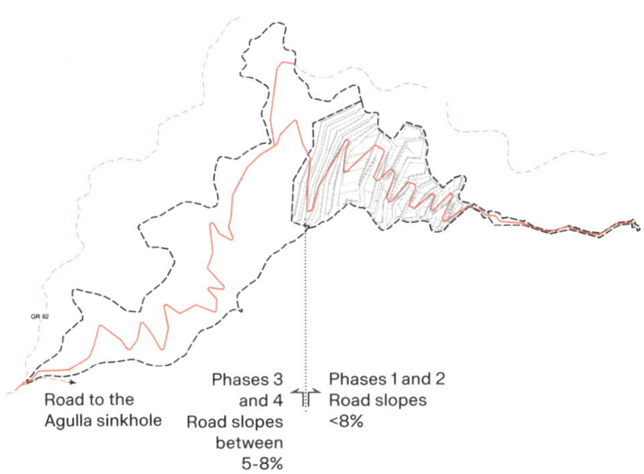

Roads and connections

Revegetation of slopes and ditches

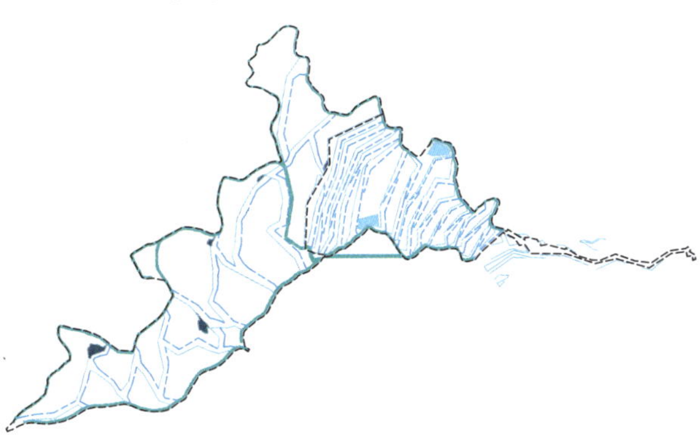

Drainage

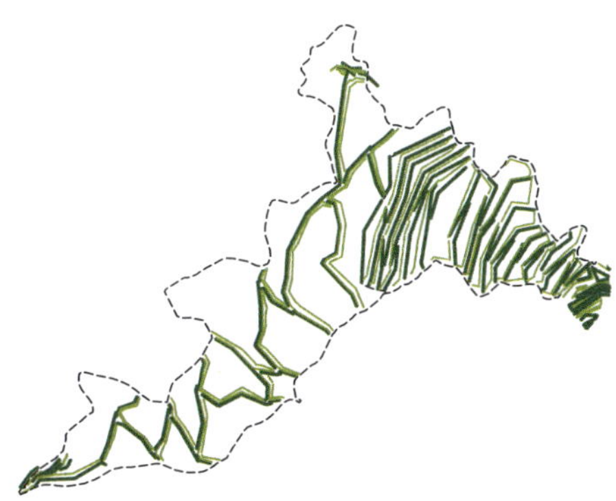

Formation of open meadows

Terrace system (phases 1 and 2)
A farming system made up of a system of terraces and slopes, where species that form part of the area's agricultural heritage were grown, facilitated the restoration activity of the first phases, creating the environmental conditions needed for the gradual invasion by the surrounding biological communities. The configuration of the various terraces required a significant contribution of outside soil which, placed on top of the waste, defined the new topography of the site.

Ridge system (phases 3 and 4)
Subsequently, in the last phases it was decided to continue implementing an agricultural system with the same characteristics yet evolving towards a system of ridges better suited to the topography of these areas, while simultaneously reducing the amount of imported soil. The result was a significant reduction in the construction budget for these works. The new system facilitates the recovery of certain Garraf bird species whose ideal habitat is an open, gently sloping environment.

Biodiversity

Slope system

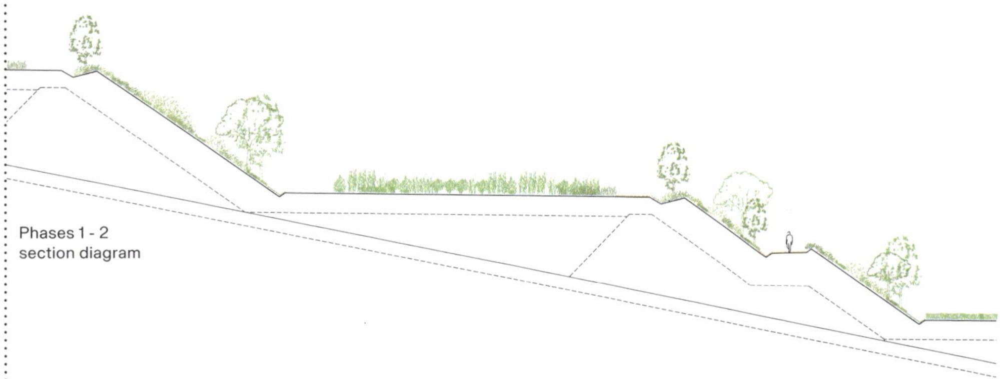

Phases 1 - 2
section diagram

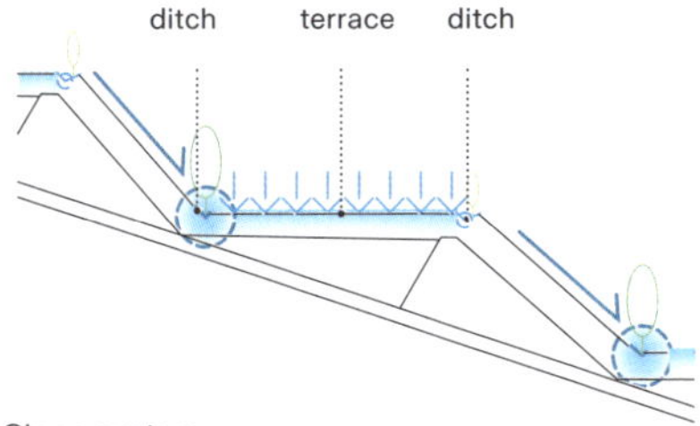

Slope system
Drainage scheme

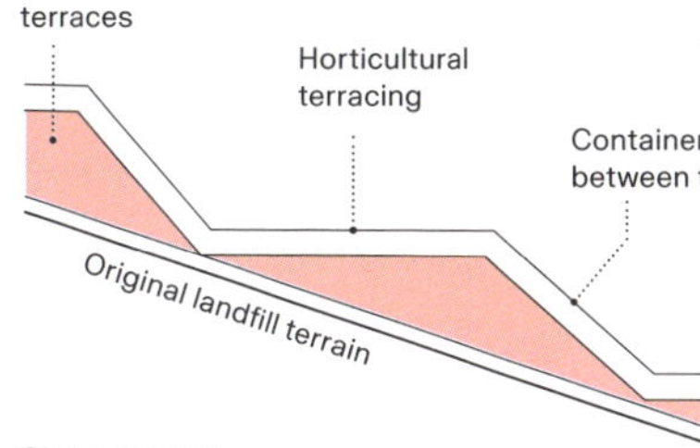

Slope system
Earthwork diagram

Ridge system

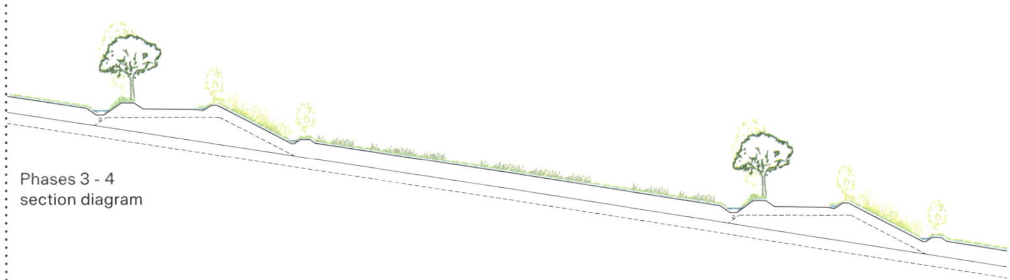

Phases 3 - 4
section diagram

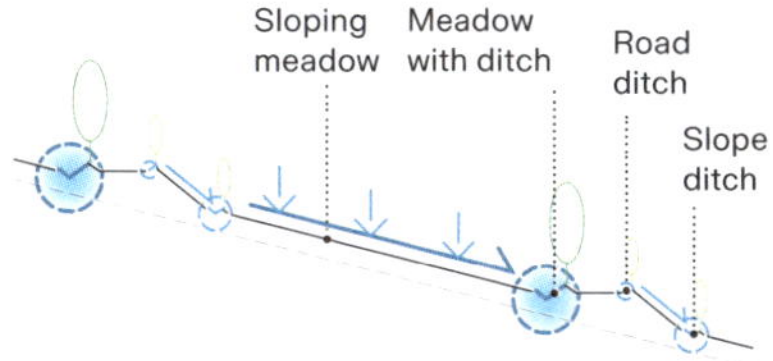

Slope system
Drainage diagram

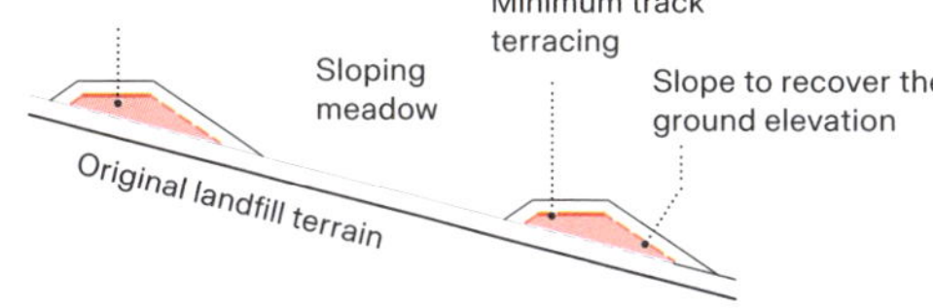

Slope system
Earthwork scheme

Typical section evolution

Innovations in the decommissioning and restoration system
During the long decommissioning and restoration process, a series of innovations were introduced into the sealed system with the aim of optimizing resources, funds and time.

The placement of geodrains for the flow of water and gas has replaced a total of 275,000 m^3 of gravel, avoiding 11,000 truck journeys to the site.
Regarding the topsoil layer, an experimental plant topsoil production process was conducted by creating technosoil with properties suited to the superficial substrate of the new meadows. Soil from outside, very poor in organic matter and of inferior condition, was mixed with digest, a subproduct generated in ecoparks from the organic material of brown and yellow recycling bins. The resulting soil is well-structured and contains a substantial amount of organic material, ideal for the development of the crops grown.

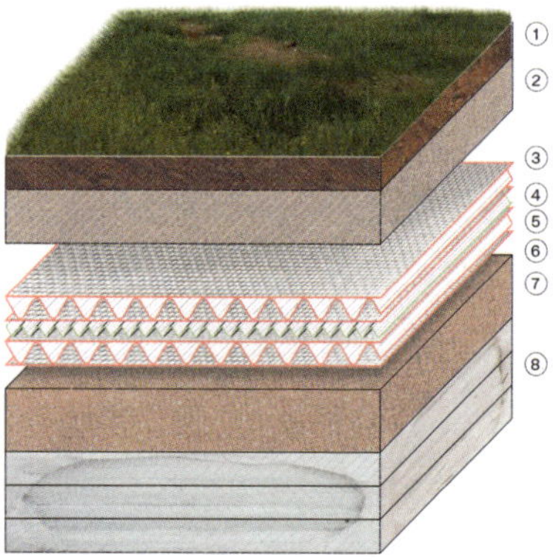

Terrain section

1 / Surface planting layer (30 cm) by technosoil formation
2 / Selected soil without compaction
3 / Geodrain. Water drainage layer
4 / Waterproofing sheet (2 mm)
5 / Geodrain. Gas drainage layer
6 / Preexisting situation
7 / Presealing. Layer of settlement soil
8 / Last waste layers

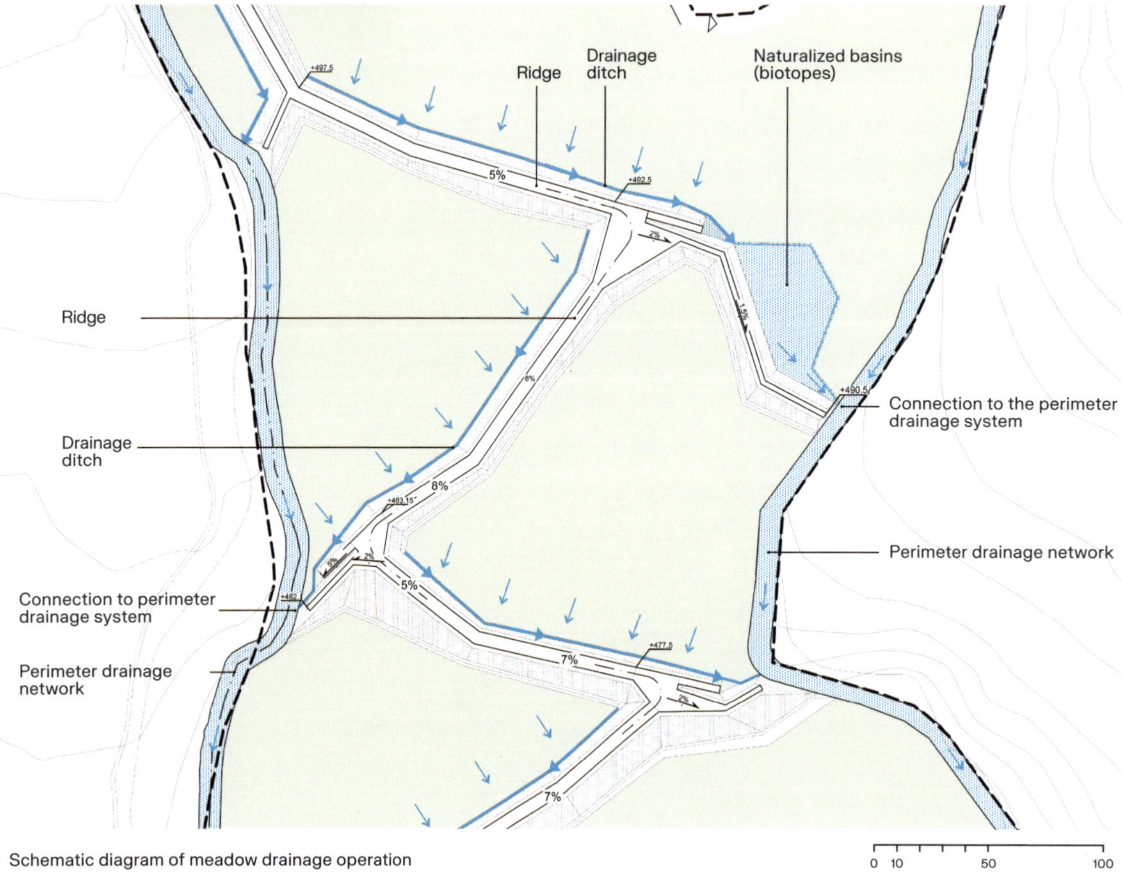

Schematic diagram of meadow drainage operation

The proposal benefits from the new drainage system using meadows to create a series of naturalized ponds. The vast areas of the resulting meadows with their unique orography allow for the creation of low points to accumulate water and foster these new ecosystems. Their importance is crucial to encourage biodiversity and as a water source to attract and support local fauna.

Phases 1 and 2

Phases 3 and 4

New amphibians colonizing the ponds

Nesting of birds
The lowlying, open meadows in areas 3 and 4 prior to the decommissioning and restoration were ideal for the nesting of many bird species rarely spotted now in Garraf Natural Park and as a hunting area for Bonnelli's eagle. These conditions, considered during the planning of construction work, led to various changes in design and choice of plant species to benefit their habitats and provide them with food.

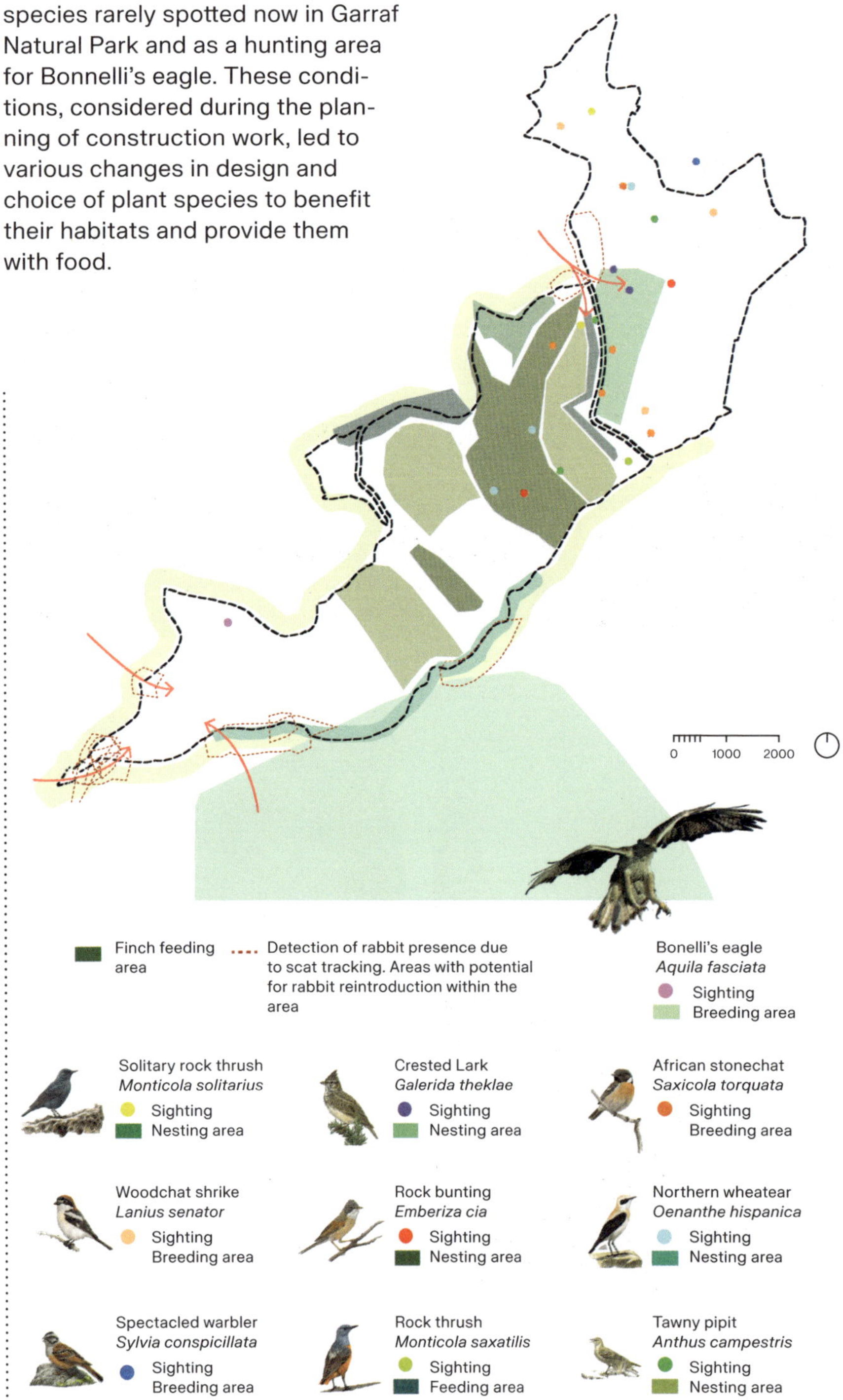

Environmental sensibility

Construction of warrens
Finally, crops and warrens were incorporated in the areas where rabbits are present, to foster their implementation and reproduction and thereby improve the condition of this space as a hunting area for Bonnelli's eagle.

Warrens

Green walls were built in the entrance area of phases 1 and 2, and were complemented by plastic waste walls that mark the gateway and remind us of the history of the place. Over the years, nature has taken them over, transforming them into hotels for insects and other species that increase the biodiversity of the place.

PHASES 1 & 2. Client AMB – Àrea Metropolitana de Barcelona / **Authors** Enric Batlle, Joan Roig, Teresa Galí-Izard / **Team** Jordi Nebot, Xavier Ramoneda, Mario Súñer, Dolors Feu, Elena Mostazo / **Collaborators** Proser Projecte i Serveis; RDS; Geocisa – Engineering / **Contractors** Urbaser, FCC, Cespa, Comsa i EMTE / **Surface area** 30 ha // **PHASES 3 & 4. Client** AMB – Àrea Metropolitana de Barcelona / **Authors** Enric Batlle, Joan Roig, Iván Sánchez, Mario Súñer / **Team** Clara Terradellas, Yago Cavaller, Lidia Carrillo, Elisabeth Torregrossa, Laura Luque / **Collaborators** SBS Simón i Blanco – Civil Engineering / **Contractor** TIRSSA / **Surface area** 45 ha

Biophilic Buildings

Biophilia is the innate connection between people and nature. Strategies that articulate relationships between people, architecture and nature drive the design of biophilic buildings that promote the biodiversity of their environment, reclaim the nature destroyed in our cities and promote the well-being of their users.

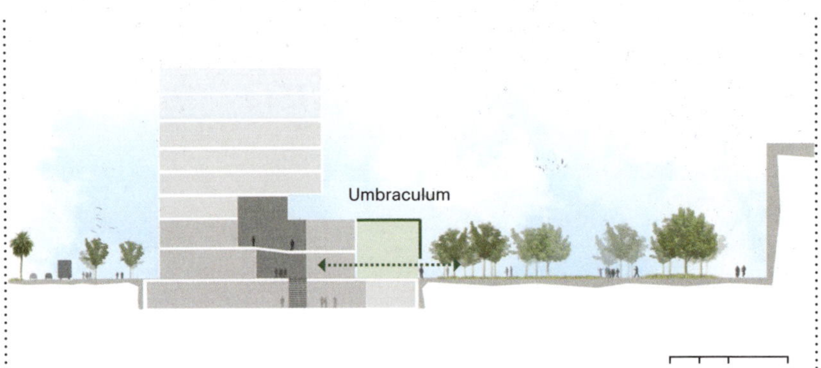

The shade structure for the southern façade of the **Illa Diagonal Shopping Center in Barcelona**, currently in the design phase, is an example of how to create biophilic spaces that encourage user stimuli.

Designing spaces in contact with nature

Incorporating outdoor terraces in buildings with vegetation or other natural elements such as water and spaces where users can experience thermal variations or air currents encourages the connection of people with nature, generating favorable biophilic reactions.

Building with natural elements

We not only appreciate nature through our eyes. Auditory, tactile and olfactory stimuli also contribute to biophilia. Building with natural elements such as wood, allowing users to touch it, also enhances this bond.

Generating spaces that encourage user stimuli

It is important to create the necessary conditions to trigger people's awareness of their changing environment, the variation of natural processes such as change in vegetation, season and time of day. We need to create conditions that support the variation of light and shadow, facilitating our perception of circadian rhythms throughout the day.

Parc Glòries Office Building in Barcelona (Spain, 2014-2019): a highly-efficient ceramic and plant façade that encourages biodiversity and extends throughout the Ca l'Aranyó gardens.

825 m of planters for a green enclosure facing the park

57,600 m of ceramic slats

610 m² of terraces

LEED
PLATINUM
Certification

BREEAM
Certification

Parc Glòries is an office building in Barcelona's 22@ technological district. It has a longitudinal floor plan with a façade on Carrer Ciutat de Granada with 17 floors. The ground floor has a veranda that extends along the rear façade overlooking the park, and the main entrance to the offices is located on the chamfered corner. Floors 1 to 8 occupy the entire length of the building, while floors 9 to 16 take up nearly three-quarters of its length, setting back the front end of the space facing Carrer Tànger. Both have open floor plans with no pillars, with a compact communication nucleus and outdoor emergency staircases at both ends.

Barcelona's Eixample. The building recognizes Barcelona's urban expansion district and adds to two chamfered corners on Carrer Bolívia and Tànger.

The basic action criteria focus on three main areas:

1 / Savings: material, energy and water resources.

2 / Environmental quality: minimization of external and internal pollutants, quality of spaces in terms of natural lighting, material comfort and air quality.

3 / Providing the building with the necessary equipment to promote the health of its users and sustainable use and contribute to the environment: terraces and biophilic spaces, spaces that encourage recycling, facilities to promote sport, physical health and rest areas, in contact with nature and fresh air.

Plans

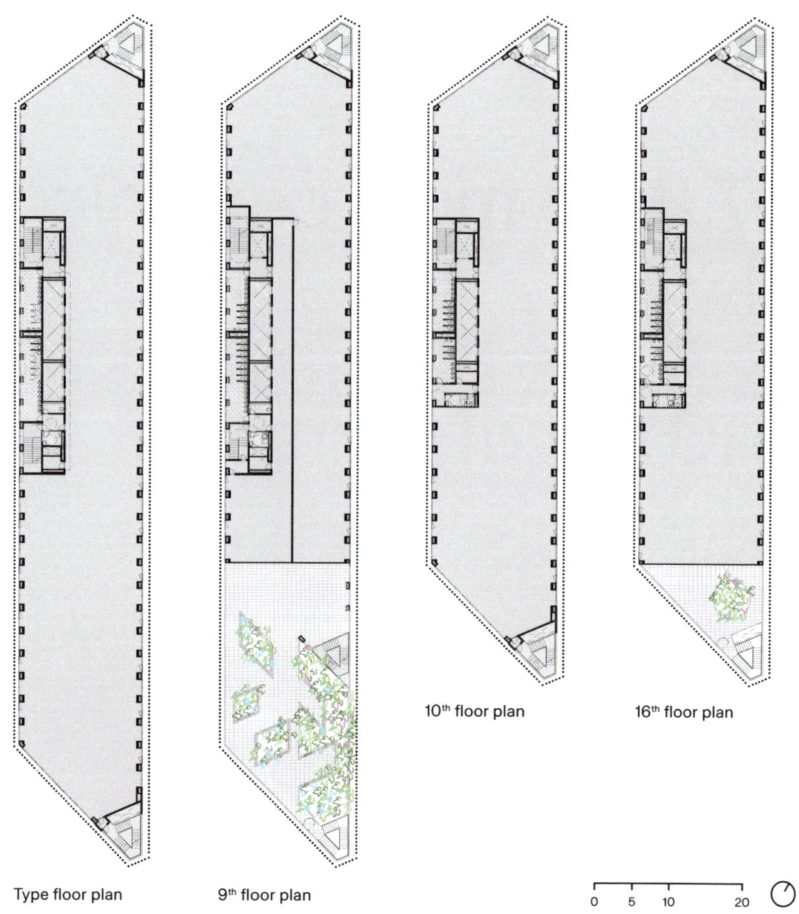

The Parc Glòries Office Building is a biophilic building that interacts with the Ca l'Aranyó gardens in the inner courtyard of the block. It seeks an ecological connection with nature, improving air quality and strengthening biodiversity. The building enables natural light to enter all its spaces and offers direct visual contact with vegetation on terraces and in planters that flood the rear of the building with greenery.

Cross section

Building

The building's energy consumption has been streamlined through the meticulous design of its enclosures. This includes its orientation and location, and incorporates logical sustainability criteria:

1 / It recognizes the solar incidence angle on each of its façades, while the **double skin** is designed with a varying degree of density depending on the direction the façade faces, filtering solar incidence and allowing maximum use of natural light.

2 / The building **framework** is in the façade, reducing the glassed surface of the enclosure by 50 % and creating a completely open-plan office floor with no intermediate pillars.

3 / The service **cores** are located on the south façade, reducing the amount of direct sunlight in the office area.

4 / Opening window and door frames encourage natural ventilation.

5 / The **roof** is designed to avoid the heat island effect by using materials that reflect sunlight and have minimum thermal inertia.

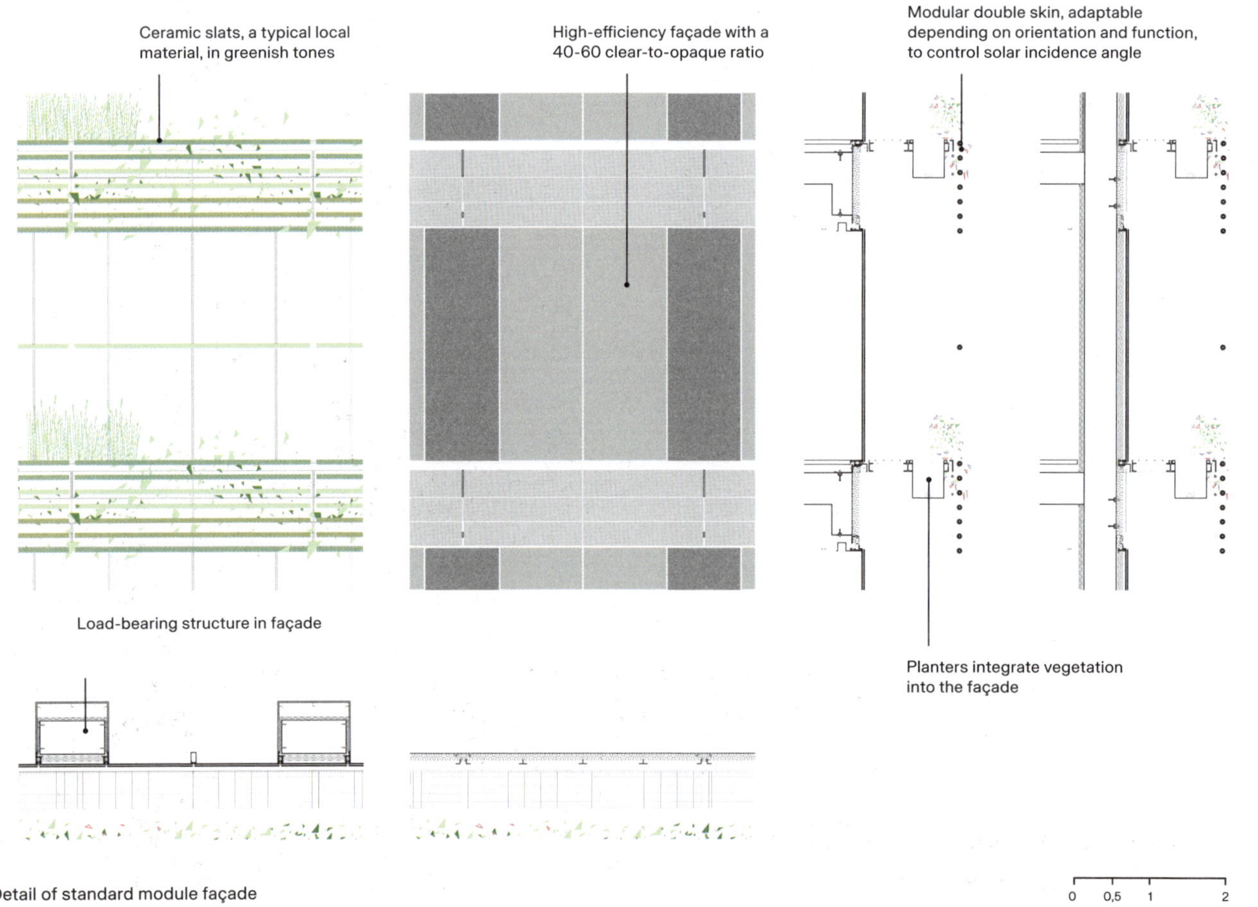

Detail of standard module façade

Industrial tradition in Poblenou. The project acknowledges the history of the place and its industrial tradition, incorporating ceramic as the primary material for the façade and the green typical of *Modernisme* to evoke traditional ceramic pieces.

User health is closely linked to the building and interaction with the natural surroundings. The Parc Glòries Office building recognizes circadian rhythms, allowing natural light to enter 100% of the spaces, including bathrooms. It also facilitates direct visual contact with natural vegetation on terraces and in planters that flood the rear of the building with greenery. A biophilic building not only contributes to the well-being and mood of its users but also helps to improve their productivity and performance.

Client Moorage Inversiones 2014 / **Authors** Enric Batlle, Joan Roig, Albert Gil / **Team** Miriam Aranda, Annabel Barba, Maria Beni, Maria Duran / **Collaborators** STATIC – Structural engineering, G3 – Building Engineering, ARCO – Façade consultancy, JG Ingenieros – Facilities engineering / **Contractors** VOPI4, Grupo Eiffage / **Gross floor area** 30,733 m²

Making the Water Cycle Visible

Cities today are specifically designed to conceal the processes that support life. When it rains, the water courses quickly to the gutters and immediately disappears into the sewage system, entering, together with the gray water from buildings, a subterranean network. Through this network – assuming it is streamlined – water travels several miles to the indispensable wastewater treatment plant, which will allow it to be reincorporated into the natural cycles.

But could it be different? Cities can and should respect the original geography, promote the conservation of small watercourses and encourage the visibility of all rainwater collection and drainage systems. Water can be collected in buildings, retained and infiltrated in public spaces and conveyed to nearby natural spaces that we can reclaim. This creates a territorial drainage system that rediscovers the continuity of water, encourages the creation of new ecosystems and facilitates the implementation of new civic continuities.

Water

Original Geography

SUDS - Sustainable Urban Drainage Systems

100% Reuse

Original Geography

The recognition of original geography and, in particular, its main element, watercourses, is essential for structuring the territory. Rivers and streams threatened in the past should be seen today as opportunities to create axes of urban development, recovering them as parks and green corridors, spaces with central civic and ecological importance that become important parts in the green infrastructure within the metropolis.

Vaguada de las Llamas Atlantic Park in Santander follows a preexisting stream that had formed several wetlands – reed beds – at its center. The initial municipal approach was to start from nothing and proceed with the construction of the park. Meanwhile, environmental groups tried to block the development of the park to preserve the reed beds. Our proposal envisions a park that respects the form of the watercourse and preserves in its central part all the preexisting wetlands converted into the fundamental rationale for the site – the Atlantic – from which all the other strategies can be explained. The result has been to the liking of both parties, and the site has become a bird reserve in the city center.

Strengthening water as a backbone of the territory

Water courses, which form the basic structure where human settlements are established, shape the natural geography. A respectful urban development should maintain these water routes as veritable green infrastructures that extend to the whole territory, preserving its natural characteristics and biodiversity.

Avoiding the typical *tabula rasa* approach of contemporary urban planning

When designing our cities, we must learn from traditional urban planning and integrate the original geography of the territory: topography, watercourses, agricultural traces and historical elements. This reduces economic costs, avoids the effects of extreme climate phenomena and maintains the inhabitants' identification with their past.

Turning geographical irregularities into a central civic and ecological space

Responsible urban planning should not reject the unique elements of the geography – mountains, rivers, riverbanks, wetlands, beaches – but preserve them, turning them into central spaces in the metropolis where citizens encounter each other and nature.

Riera de Sant Climent Linear Park in Viladecans, Barcelona (Spain, 1996-2008): an urban and natural corridor connecting Baix Llobregat agricultural park and the Garraf massif through the town of Viladecans.

80% maintenance of the open stream

2 km uninterrupted linear park

100% ecological and social continuity between El Garraf and the Llobregat Delta

The municipal interest in reclaiming the Sant Climent stream became a framework plan that organized all the projects undertaken along its course. The advanced planning did not foresee the possibility of creating this linear park, opting for open spaces that were fragmented and situated according to different urban paradigms. The result is an example of how ideas of this kind can be implemented while maintaining all other urban planning parameters. The park design follows the logic of the stream's natural system. Making use of all the spaces it can, it becomes the backbone of the town, connecting the mountain (El Garraf) to the sea (the delta).

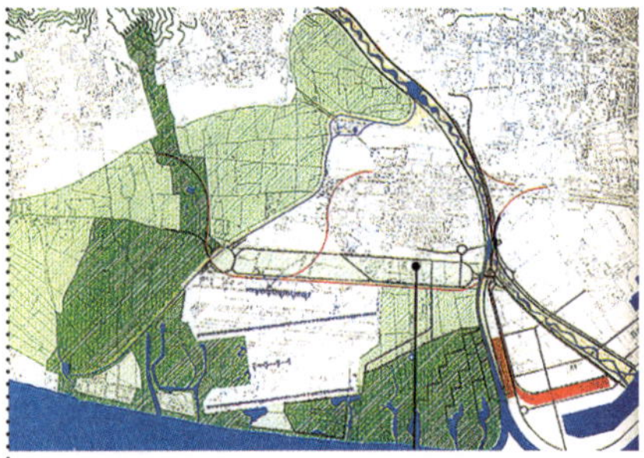

From satellite to magnifying glass
In one of the many proposals to expand Barcelona's airport, the environmental group DEPANA proposed a green corridor between El Garraf and the delta between Viladecans and Sant Boi de Llobregat.
The corridor never came to fruition because the preexisting industrial parks were expanded instead. Years later, however, based on a more detailed analysis, we saw that the original geographical element endured: the stream.

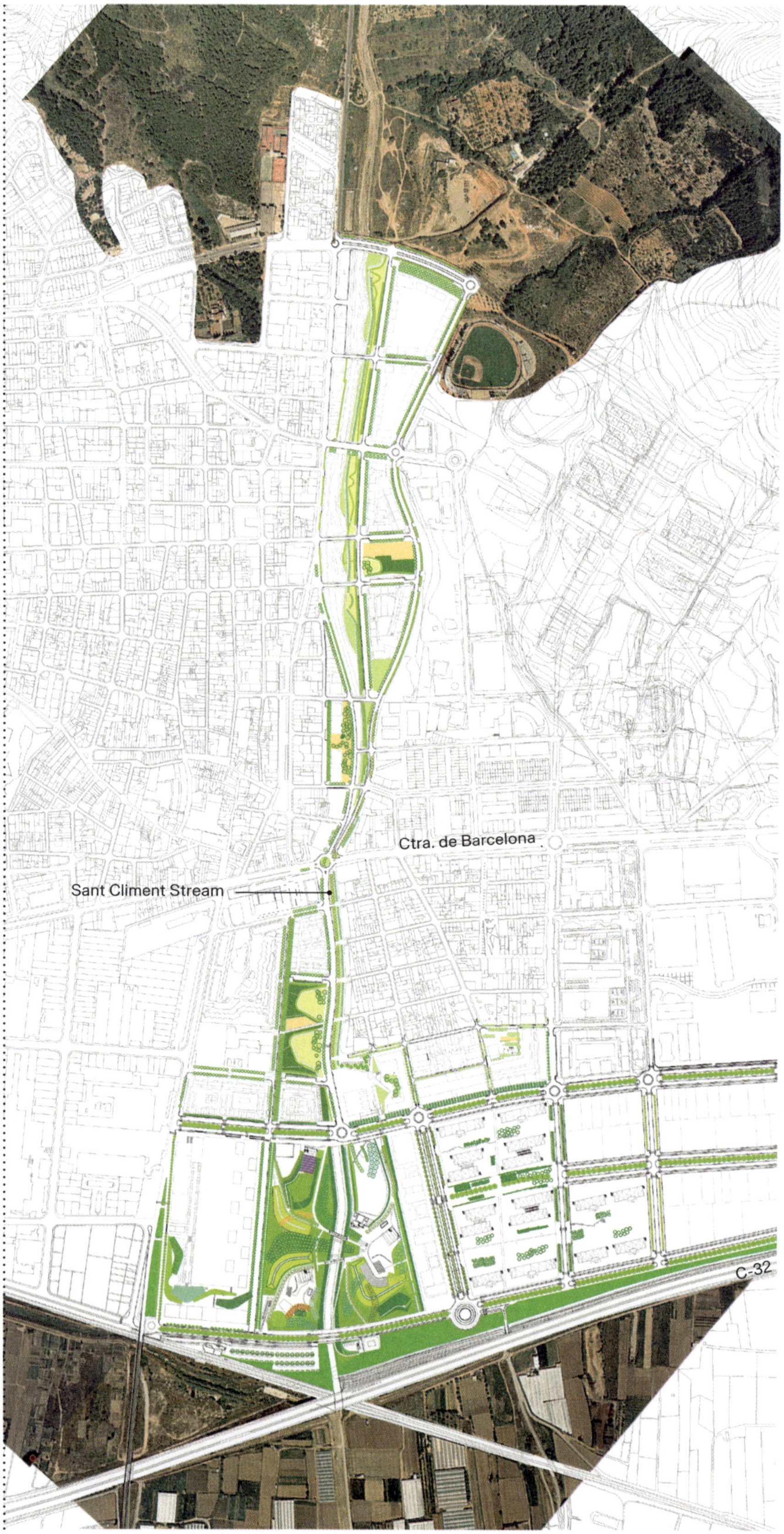

General plan

Perhaps the stream was not wide enough to be considered a green corridor, but it is the geographic element that connects El Garraf to the delta. Based on detailed work in each section, the planned urban developments, which nearly eliminated the creek, were transformed in the city's most ambitious undertaking: a system of open spaces that organize and connect the entire area.

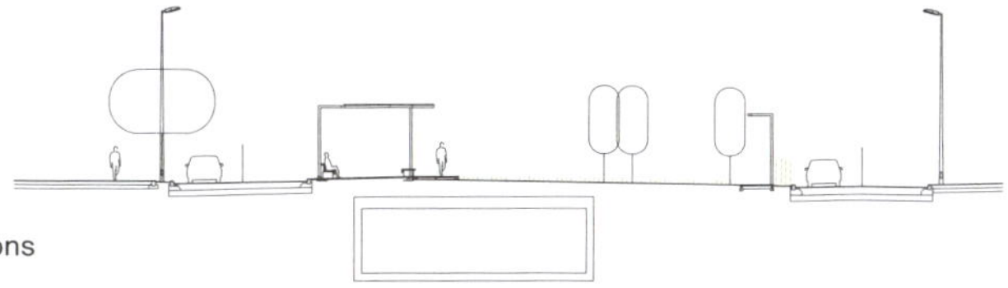

Buried sections

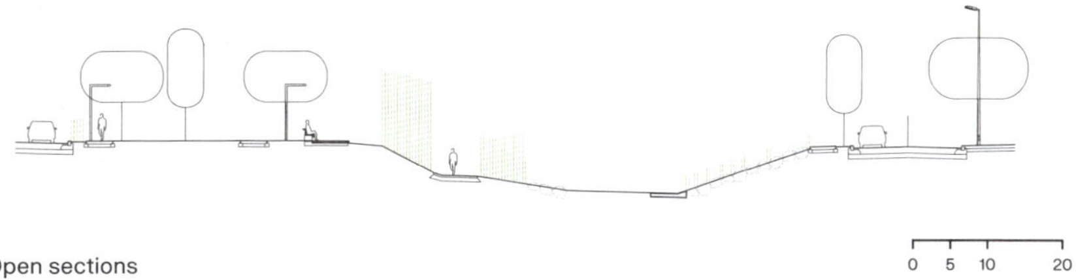

Open sections

Parc de la Marina (2003-2008) is the culmination of the Riera de Sant Climent green corridor. The stream flows continuously, creating basins that, in case of floods, maximize rainwater infiltration to the subsoil and use of rainwater as natural irrigation.

RIERA DE SANT CLIMENT LINEAR PARK IN VILADECANS. Client Viladecans Town Council / **Authors** Enric Batlle, Joan Roig / **Team** Jordi Nebot, Xavier Ramoneda, Mario Súñer, Diana Calicó, Teresa Galí-Izard, Elena Mostazo / **Collaborators** STATIC – Structural Engineering / **Contractors** Proinosa, Copisa, Comsa, 4CT / **Project area** 160,000 m² // **PARC DE LA MARINA. Client** Viladecans Town Council / **Authors** Enric Batlle, Joan Roig / **Team** Xavier Ramoneda, Mario Súñer, Elena Mostazo, Diana Calicó / **Collaborators** STATIC – Structural engineering / **Contractor** Proinosa / **Project area** 96,850 m²

SUDS - Sustainable Urban Drainage Systems

Cities today are ecosystems with an inert, impermeable skin. Cities need a new permeable, biophilic skin that acts as a thermal insulator and as a water and atmospheric filter to recover the biorenewable capacity that is currently lacking. New designs should incorporate strategies to maximize soil permeability, infiltration and rainwater retention in the public space. SUDS reduce runoff and network saturation, improving landscape and urban environmental integration.

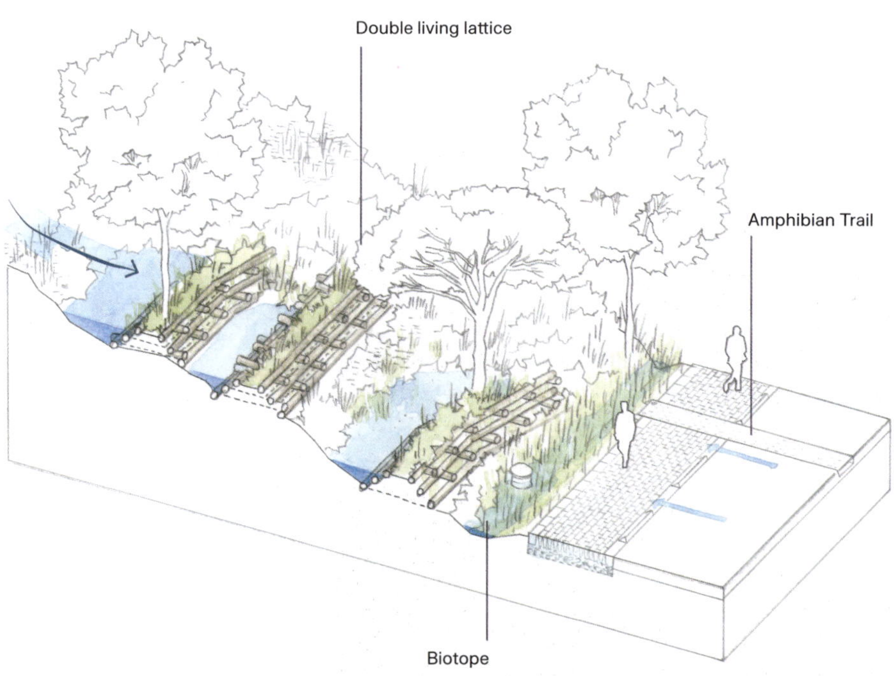

The proposal for **the Parc dels Tres Turons Tender in Barcelona** established a decalogue of bioengineering-based solutions for the recovery of the edaphological quality of the soil by improving the management of surface runoff and the use of water by absorption: live stakes, bundles, coconut fiber plant structures, forest trays, as well as lattices and Krainer walls for areas with steeper slopes. Rainwater retention dikes help to prevent erosion. The proposed system transforms the problem into a value.

Recovering the natural water cycle

We need to reconsider fresh water as a source of life by integrating it into public space projects without wasting it, to reduce consumption and the dependence of cities on the rest of the territory. We must reestablish the natural water cycle in open spaces to make our urban drainage network more rational, simplifying construction solutions and facilitating infiltration in green areas.

Maximizing soil permeability to create a "sponge" city

Maximizing permeable surfaces in public spaces will help promote the recharging of water tables with rainwater. Keeping the soil strata hydrated by creating low wetlands protected from direct solar radiation will ensure natural hygrothermal regulation of the substrate and encourage evapotranspiration in landscaped areas.

Promoting rainwater harvesting and management

We need to design our public spaces to support the collection and accumulation of water during rainy periods and its subsequent reuse for irrigation of green areas during dry periods. Defining an orography of open spaces compatible with natural surface runoff will improve water management and facilitate the maintenance of public space.

MEETT Exhibition Park in Toulouse
(France, 2011-2020): a landscape architecture undertaking inspired by agricultural tradition and designed with water in mind that seeks to integrate this large facility into its environment and restore the water balance in a self-sufficient way.

+20,220 m^3 water retention in ponds and ditches

100% rainwater sustainably managed within the sector

100% land use

The implementation of new buildings in the exhibition park designed by OMA and the infrastructure they require entails a substantial alteration of the surrounding landscape and a high degree of surface waterproofing. The elevated water table level, at least 1 meter deep, exacerbates this already complex situation in the area close to the Garonne River.

The landscape architecture project seeks to integrate this major strategic complex within the surrounding agricultural mosaic through different techniques and the communication systems of cultivated fields. The project also takes on the challenge of reestablishing the water balance of the place, interrupted by newly built waterproof surfaces.

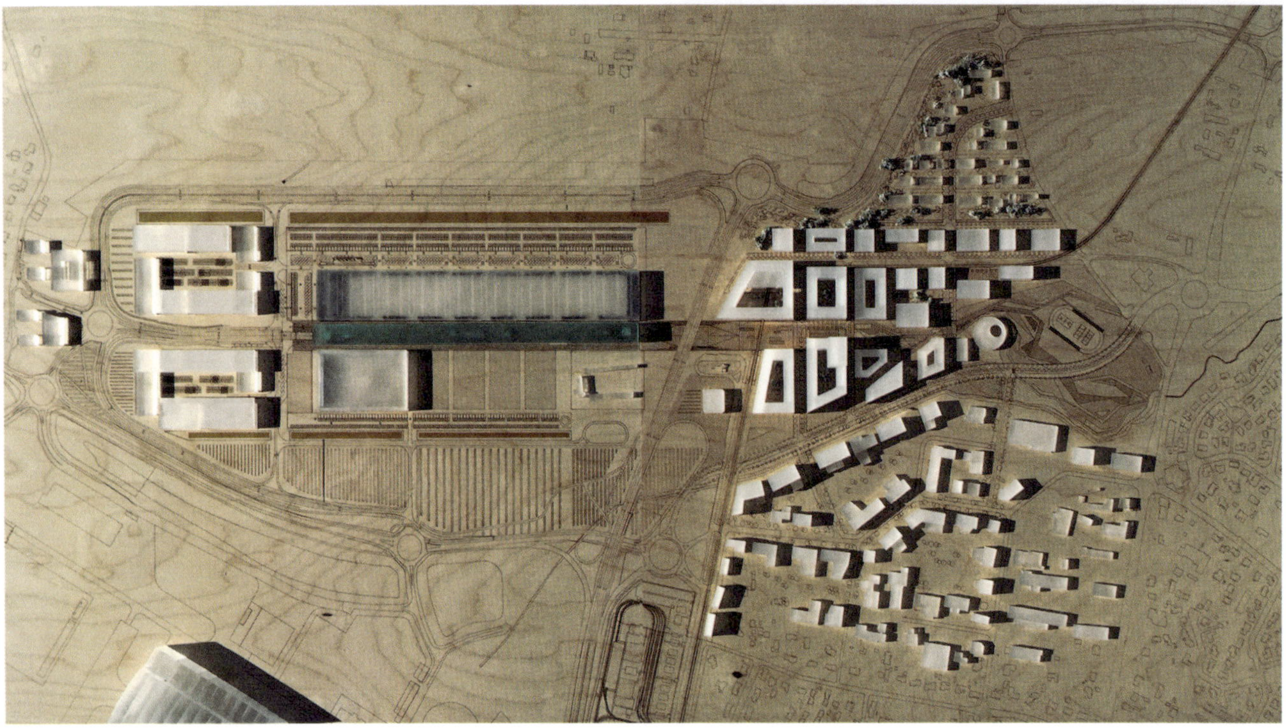

Model of the proposal

A system of solutions is proposed to absorb the water of this runoff, retain it and filter out contaminating particles from nearby roads or left on roofs by atmospheric pollution, through gradual phytodepuration processes.

1 / Farm furrows

2 / Tree ridges

3 / Lagoon ponds

Explanatory illustrations of basic project tools

1 / Farm furrows
The linear forms of furrows in nearby farmland are reproduced and subtly inserted into the open spaces of the complex. The resulting patterns define the new landscape of the surroundings with a familiar language and generate a rainwater interception and infiltration system. The vegetation is varied, consisting of lowland Mediterranean, continental and Atlantic forests, in keeping with the climate in Toulouse.
These green structures create ecosystems where the morphology and the seasonal presence of groundwater give rise to ideal conditions for the spontaneous development of plant life. They are conceived as lifelines in the landscape capable of taking advantage of the natural dynamics of the site itself.

2 / Tree ridges
Also based on the habitual patterns of the agroforestry mosaic at the site, trees are planted on ridges. These provide optimal humidity and ventilation conditions for the root system of the trees where the groundwater level is very high.
The creation of this system of ridges helps to achieve neutral balance in the management of the earthworks required to carry out the proposal. It also is an ideal planting method, raising trees above the water table.

3 / Lagoon ponds
Runoff water terminates in large retention ponds and basins where it is absorbed and filtered by phytodepuration, before it is carried by infiltration to the Garonne riverbed. A new green infrastructure is capable of resiliently managing large rainfall events.
It is a landscape in continuous movement that, through the fluctuations in water levels and seasonal changes, fosters an explosion of life and high species variability throughout the year.

1

2

3

Client SPLA Toulouse Métropole / **Landscape designers** Enric Batlle, Joan Roig, Ivan Sánchez / **Landscape team** Guillem Matallanas, Mario Súñer, Javier Zaldívar, Simon Gabillard, Dolors Feu / **Architecture authors** OMA, Puig Pujol Architecture, Taillandier Architectes Associés / **Collaborators** Ingerop – Engineering / **Project area** 190,000 m²

100% Reuse

Water is life: the new water culture consists of using local water, consuming as little of it as possible in an efficient way, reusing it as often as possible and returning it to the environment in the highest quality.

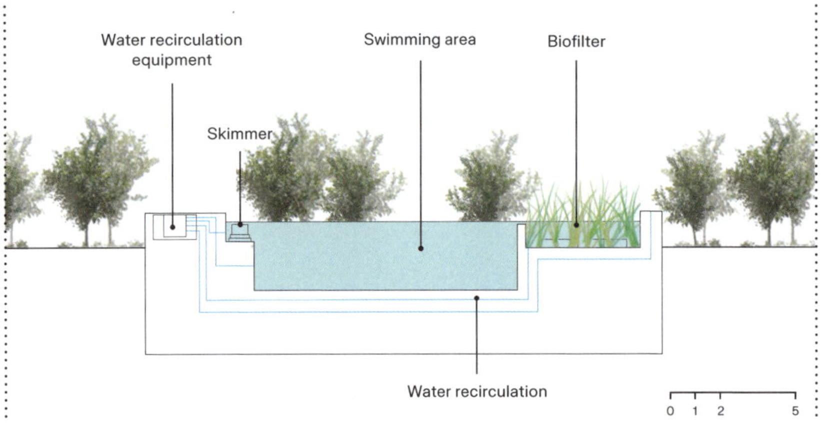

In the **Renaturalized Pool in Alcanar**, the water is kept clean and clear without the use of chemical products thanks to its constant recirculation through a biofilter with native aquatic plants that simultaneously acts as a biotope which promotes environmental biodiversity.

Reusing 100% of rainwater, avoiding its discharge into the network

Accumulating and using the water collected on the roofs of buildings is imperative. Rainwater is suitable for irrigation and other applications that do not require drinking water. Using it in buildings helps to reduce the consumption of water and all the resources needed for its purification and distribution. In addition, the surplus can be reinfiltrated into the land itself, mimicking hydrologic behavior in natural spaces, thereby avoiding its discharge into the network.

Using reclaimed gray water as an alternative system

We can reuse water considered only slightly contaminated such as water from showers, bathtubs and sinks. Residential buildings generate enough gray water to cover their water needs for non-drinking uses (toilet flushing, cleaning and irrigation).

Implementing natural water purification systems

We need to create natural purifiers that can become gardens. An example of this is the Living Machine in which wastewater circulates through different platforms with species capable of purifying and growing. Treated water can be reused for irrigation or other non-drinking uses or returned to the environment in the highest quality.

The Style Outlets in Viladecans, Barcelona (Spain, 2011-2016): a contemporary, warm, natural architecture defines a metropolitan and urban area where the topography, nature and water take center stage.

0 liters of rainwater discharged into the sewage system

100% use of rainwater

66% rainwater infiltration into the soil

BREEAM
Certification

The Style Outlets shopping center, which sits on a nearly 7-hectare trapezoid-shaped plot of land in the municipality of Viladecans, simultaneously adapts to two widely differing situations. On the one hand, it conforms to the metropolitan landscape with direct access from the C-32 freeway, located at the lowest level of the site, indicated by a large gateway visible from the highway. On the other hand, it adapts to the landscape of the town of Viladecans with an urban access located at a higher level and connected to the railway station via Carretera de la Vila road, which extends from the town to the beach.

The plot is located within the Baix Llobregat Agricultural Park in unique topographical and hydrological circumstances. We opted for a rainwater management design based on the collection of precipitation and inspired by hydrological functions in natural spaces. The system facilitates evapotranspiration thanks to the abundant vegetation, and guarantees infiltration, retention, accumulation and reuse for the irrigation of all landscaped areas. Water management guarantees 100% rainwater recovery and avoids its discharge into the network.

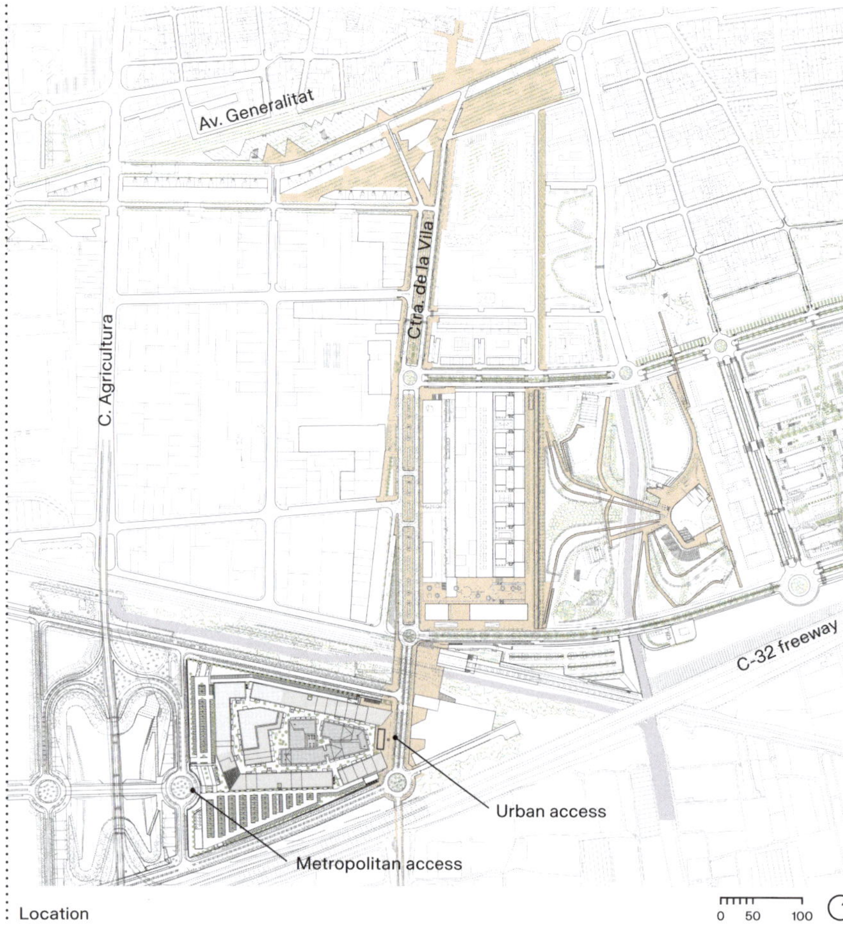

Location

The center, with over 100 stores, is organized around an open ring-shaped promenade with two main entrances that provide access to all the shops. The fragmented and variable geometry of this promenade creates a varied range of spaces among which the large central square where most of the restaurants are concentrated stands out.

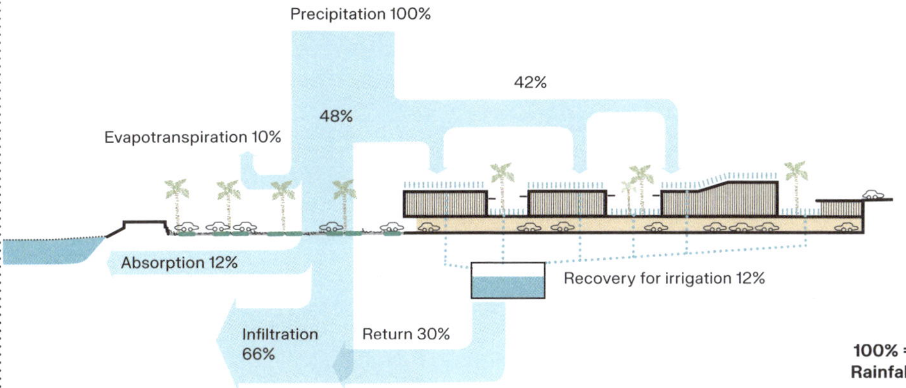

Water filtration diagram

Analysis of designed system during one day of actual rainfall: 15 l/m² of rain (average rainfall in the area):

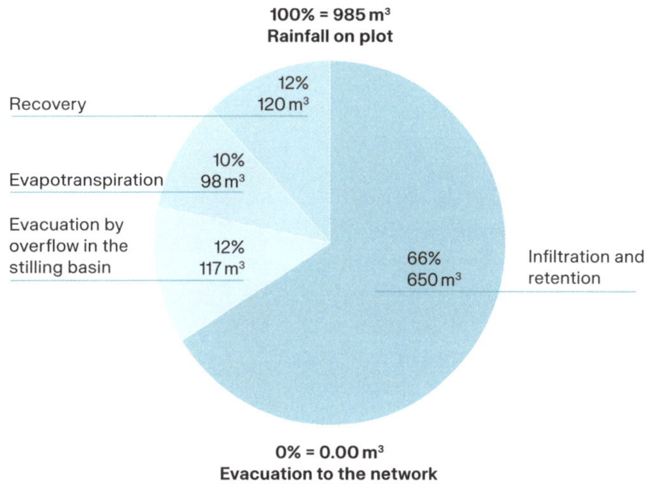

Outdoor parking retention system based on Sustainable Urban Drainage Systems (SUDS).

The design of the shopping center's topography reflects the conditions of the place, improves the functioning of the shopping center and contributes to implementing more environment-friendly solutions. A smooth topographic adaptation allows the parking lot to be developed in a unified way on a single level, enhancing natural lighting and ventilation as well as contact with nature.

The project is part of the new smart sustainability paradigm where natural materials and advanced industrial materials are combined to improve the performance of the stormwater management system and provide numerous benefits such as groundwater recharge and heat island prevention. The design of this system is the unique feature of the project.

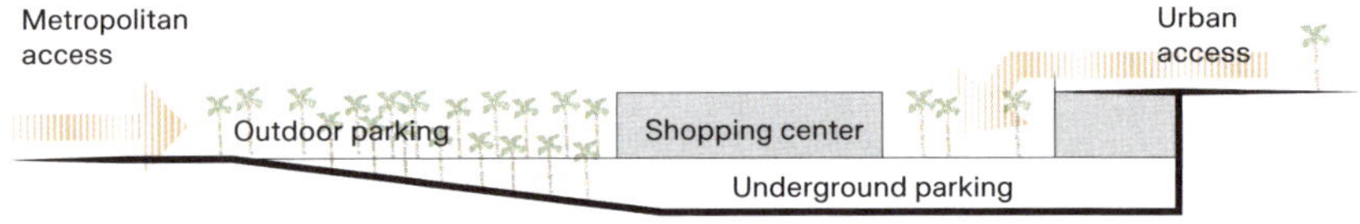

Parking lot that adapts to two topographic situations: an outdoor parking lot connects to the metropolitan access and an underground parking lot is situated beneath the commercial buildings.

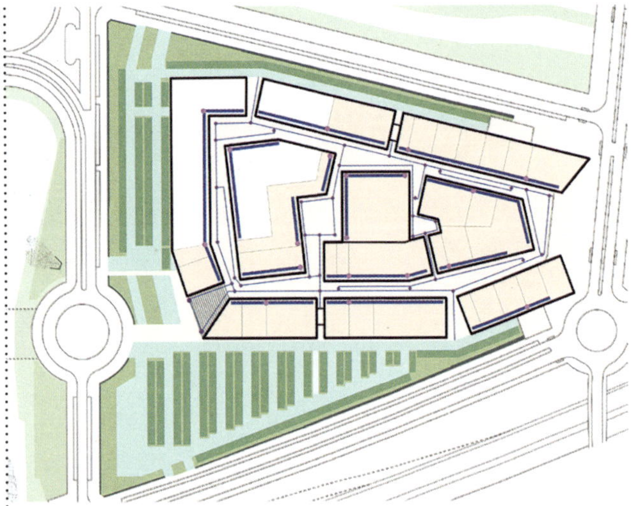

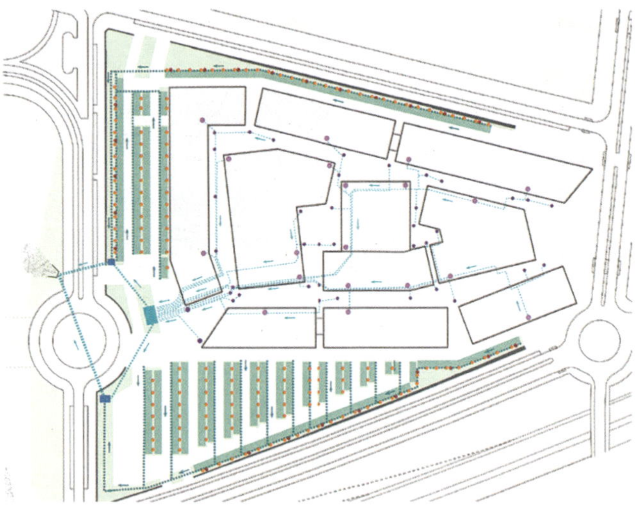

Rainwater collection schemes

Biophilia and sustainability are the pillars of this design that brings nature closer to the shopping experience and strengthens logical sustainability through the implementation of measures that guarantee maximum energy savings and reductions in consumption and CO_2 emissions.
The overall picture is of a garden that includes both perimeter parking lots and a sequence of paths and plazas at the center. The proposal is attuned to Mediterranean sensibilities and presents the image of an oasis.

Local and recyclable natural materials were used, strengthening biodiversity by planting native species that require little watering.

Client Neinver / **Authors** Enric Batlle, Joan Roig / **Team** Helena Salvadó, Eirene Presmanes, Laura Quintana, Xavier Irigoyen, Iván Sánchez, Antoni Monté, Maria Duran, Alba Puig, Diana Calicó, Elisabeth Torregrosa, Alex Gràcia, Yago Cavaller, Dolors Feu / **Collaborators** STATIC – Structural Engineering, PGI Engineering – Facilities Engineering / **Contractor** Sacyr / **Gross floor area** 50,277 m², 41,750 m² (urbanization)

Walking the City

One of the challenges of the 21st century is to connect the territory by promoting soft mobility by means of roads that facilitate ecological and social continuity between the different parts of the city. To do this, we will need to build bridges to recover lost connectivity, always seeking to promote biodiversity, increase productivity and encourage the leisure activities of urban inhabitants.

In many cases the boundaries are impassable, like walls that impede connectivity between the different parts. Overcoming these interruptions makes the project the most strategic plan possible. It means paying more attention to the difficult place rather than creating something new on one of its two sides. These places are the missing link in the chain, the passage between two fragmented situations, and resemble the military points described by Richard T. T. Forman: projects that should concentrate all our activity, because they will bring a large territory within our reach.

Sustainable Mobility

Green Routes

Door to Door

Healthy Stairs

Green Routes

We need to design a mobility system that combines the best conditions for pedestrians and cyclists based on an efficient, integrated public transport system that guarantees the highest possible quality of life. We can structure the urban fabric in relation to slow mobility routes and enhance the interchanges between these networks of routes and the public transport network as new centers of daily life, places that we must equip and densify.

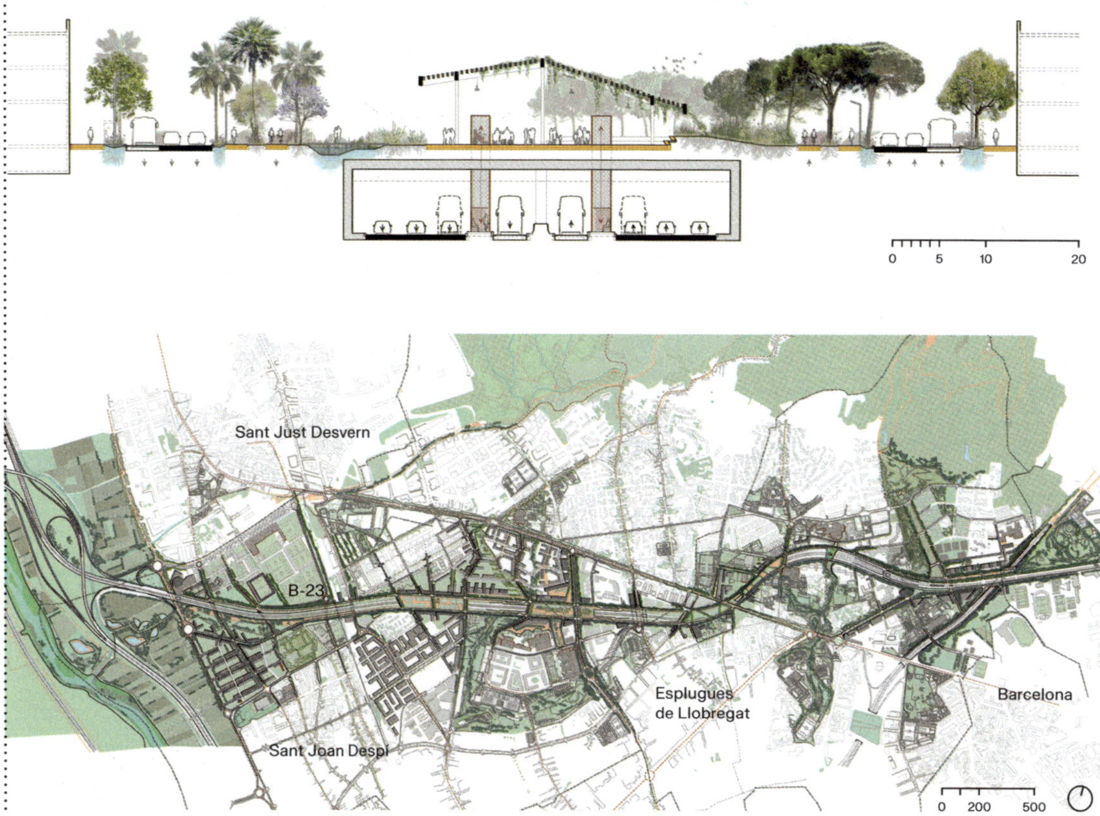

The study of the B-23 freeway between Barcelona and the Llobregat River, which we are developing in collaboration with BAU, seeks to transform what is now a barrier into an extension of the activity axis of Avinguda Diagonal, identifying possibilities for integration, connecting its sides and taking advantage of spaces that open up around it, giving special prominence to modes of sustainable mobility that are currently marginalized.

Structuring growth along the pathways of soft mobility

Just as traditional cities grew along roads, the new mobility must lead us to cities structured along slow mobility routes that are pleasant, well equipped and green, transformed into civic paths that connect with public transport nodes.

Strengthening public transport nodes as new areas of centrality

Public transport stations, often set in disorganized, poorly used environments, must be transformed into central spaces for the 21st century with the highest density of activities and mix of residential and productive uses that make the most of their connectivity.

Transforming infrastructure for private transportation

The construction of large infrastructures for private vehicles has created impassable barriers between neighboring urban fabrics. The marginal spaces that sprung up to either side must be turned into spaces of opportunity, creating areas for sustainable mobility where the urban fabrics fragmented by these infrastructures can be recovered.

Cyclable Connection Between Barcelona and Esplugues de Llobregat Along Avinguda Diagonal (Spain, 2014-2018): a new connection within the metropolitan Barcelona urban bike lanes and pedestrian network that connects Avinguda Diagonal to the system of open spaces in Collserola Park and Baix Llobregat by means of the large road infrastructure.

1,706,322* bicycles since its inauguration

(*data measured in February 2022)

880 m of bike lanes linking Barcelona and Esplugues de Llobregat

4.5 m wide for bicycles and pedestrians

The project involves the construction of a pedestrian and bike route in one of the most difficult enclaves of the metropolitan road network: the junctions of the B-20 and B-23 infrastructures, and the loop formed by the links and accesses to the two cities. This point has been a pedestrian barrier and an obstacle for cyclists for over 60 years. The bike route finally connects the two cities as directly as possible in a firm commitment to active mobility and a biophilic city that is healthier, less polluted and more accessible for people.

The new 880-meter-long passage brings the two municipalities closer together, saving almost 500 meters of travel distance. The route is easy for bicycle and pedestrian use in renaturalized surroundings, serving as a lookout over the city. Likewise, the route links the urban bike lane and pedestrian network to the system of open spaces in Collserola Park and the Carretera de les Aigües road, through green spaces in the Finestrelles neighborhood, becoming one of the shortest and most accessible approaches in the city.

Avinguda Diagonal connected on the same level with the Esplugues road and Cervantes Park in the sixties.

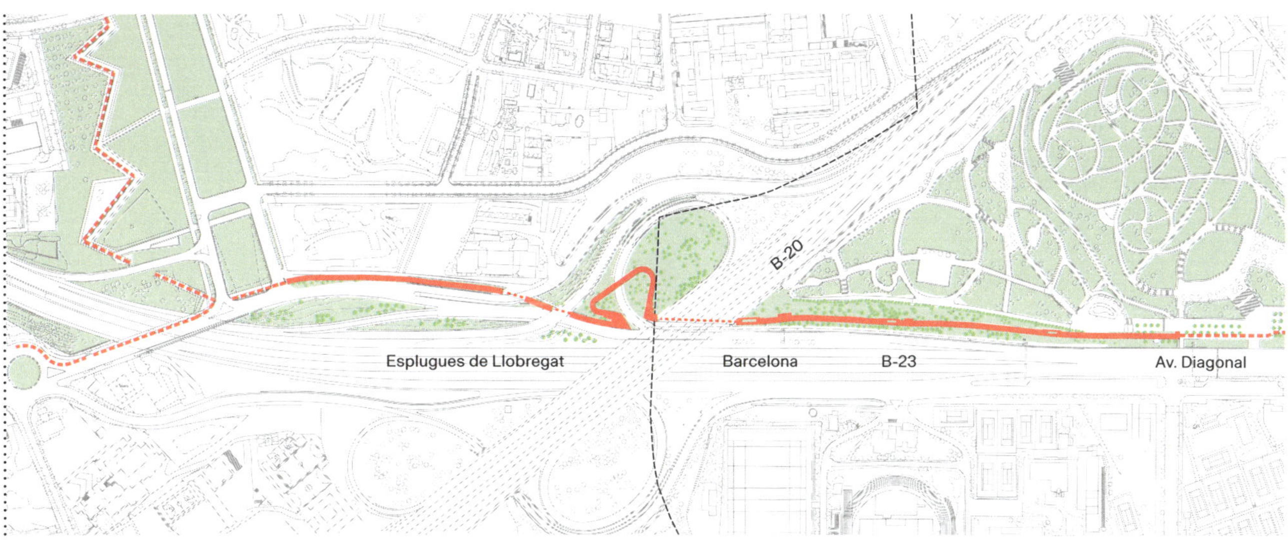

Pending the definitive solution, the Barcelona Metropolitan Area has promoted the construction of this bike lane to solve the problem. The results have exceeded all expectations.

The new route is 4.5 meters wide and accommodates two types of traffic: pedestrian and bicycle. The paving is different, both in color and texture. This enables the bike track, which is 2.3 meters wide with two traffic lanes, to maintain continuity and homogeneity along the entire route. Meanwhile, the pedestrian lane is 4.5 meters wide, gaining additional space thanks to a series of more open spaces that adapt to the geometry of the planted terrace and the existing trees. This facilitates more direct contact with nature and the creation of small rest areas with benches, seats and bike parking to enjoy views of Avinguda Diagonal.

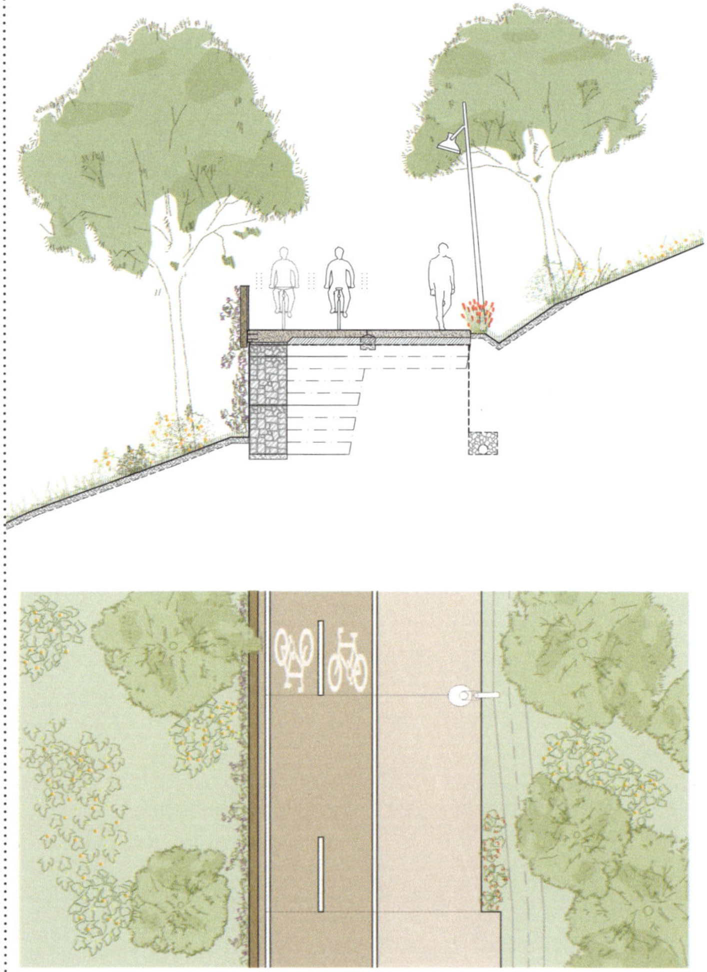

Cross section and plan of the lane

The new bike lane is a green route for calmed traffic. Located in the worst of places, it has the potential to become a pleasant, contemplative spot, evoking rural spaces away from the city and large road infrastructure. It is suitable for daily and recreational use and walking, seeking to transform a forgotten, rejected space into a meeting place that provides urban continuity to the metropolitan community.

Client AMB – Àrea Metropolitana de Barcelona, Barcelona City Council, Esplugues de Llobregat Town Council / **Authors** Enric Batlle, Joan Roig / **Team** Antoni Monté, Francesc Montero Llorca, Joan Batlle, Yago Cavaller, Dolors Feu / **Collaborators** A. Bianchini Ingeniero – Gabion Calculator, SBS Simón i Blanco – Civil Engineering / **Contractors** Rogasa, Agrotècnica del Segrià / **Project area** 880 linear meters

City and Territory

Door to Door

The promotion of environmental networks, the construction of an efficient public transportation system and the establishment of routes that allow slow mobility – pedestrians and bicycles – are three actions that can help us to hybridize our cities with each other and with the natural environments that remain. We must establish clear routes for pedestrians and bicycles with innovative designs that improve their bioclimatic conditions and overcome existing barriers.

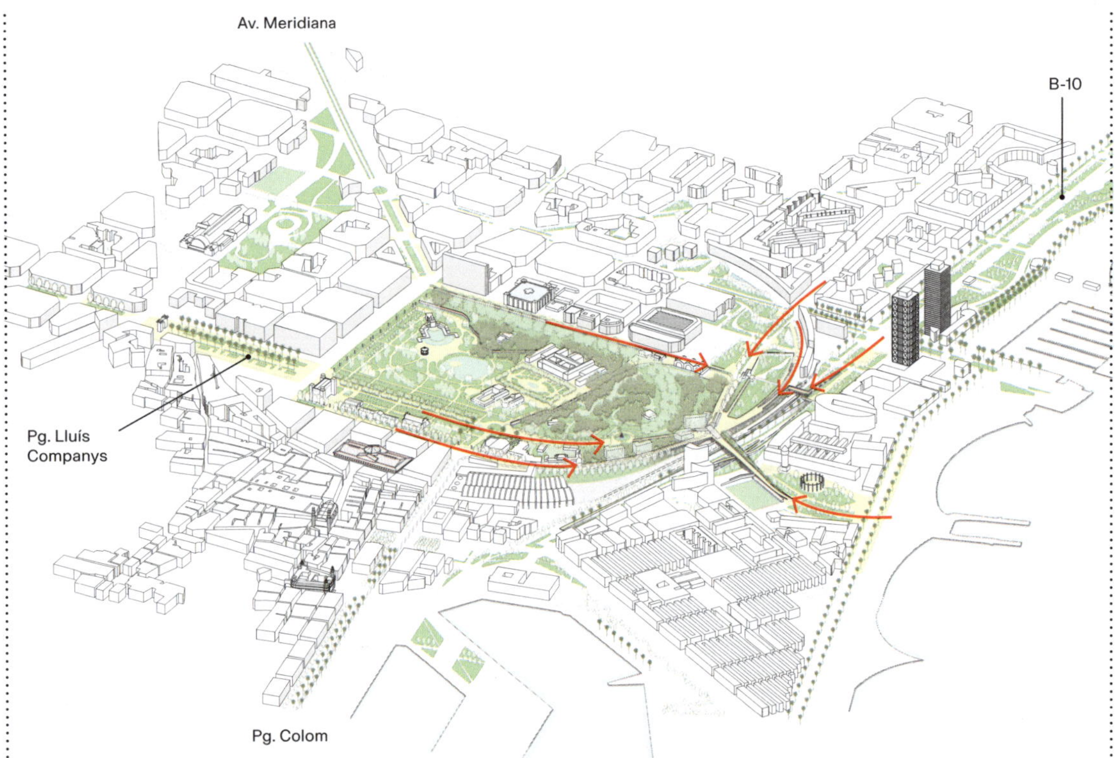

The Pasos hacia el Mar (Steps towards the Sea) study to propose new opportunities for connection between the city and the sea through Ciutadella Park is a clear example of how the city can recover the connectivity lost due to the construction of large infrastructures, introducing healthy, sustainable mobility for people. The proposal envisages an accessible path over the Ronda del Litoral ringroad and the railroad to open Ciutadella Park up to the sea and offer a direct, continuous route from Passeig Sant Joan to the seafront promenade and the beaches of Barceloneta, completing a system of open spaces that defines the city center.

Thinking of open space as a crossroads

We need to design public space for the mobility of people by incorporating existing elements of interest and linking them to give continuity to walking and cycling routes. Establishing an extensive network of paths through open spaces will enable the flow of citizens throughout the territory, in a clear commitment to sustainable active mobility.

Interconnecting open spaces

We must reactivate marginal or degraded open spaces by incorporating them into the network of approaches to urban centers. Directing the relationship of urban public spaces toward natural surroundings will ensure the continuity of the territory and facilitate direct access for people between the city and the landscape.

Overcoming barriers to sustainable mobility

Our cities are full of barriers created by traffic flows that prevent soft connectivity between separated neighborhoods. Solving these thorny issues is the most important strategic plan we can undertake, since its implementation will activate the territories located on both sides of the old barrier.

Pedestrian Connection Between Barcelona and Montcada i Reixac (Spain, 2019-2020): a safer, healthier and more comfortable connection between the two municipalities and an example of a paradigm change in urban mobility to door-to-door.

393 m route for active mobility between Barcelona and Montcada

0% topographic modification

100% conservation of existing trees

The pedestrian connection project on the outskirts of the city is located at a municipal interstice, the confluence of the Collserola and Marina hill ranges, in contact with the natural surroundings of the Besòs River and Rec Comtal.

The project aims to consolidate the narrow path that connects the neighborhoods of Vallbona and Can Sant Joan, one of the few pedestrian connections between the municipalities of Barcelona and Montcada i Reixac. The new route promotes active mobility between the two, where the route and the slope of the path on the side of Vallbona Hill are evened out and the social connectivity of the metropolis is improved, promoting contact with quality natural spaces. Despite the rugged topography of the place, after the tunnelling of the railroad tracks this accessible route will be the only direct connection for pedestrians between Montcada and Barcelona along the Besòs River.

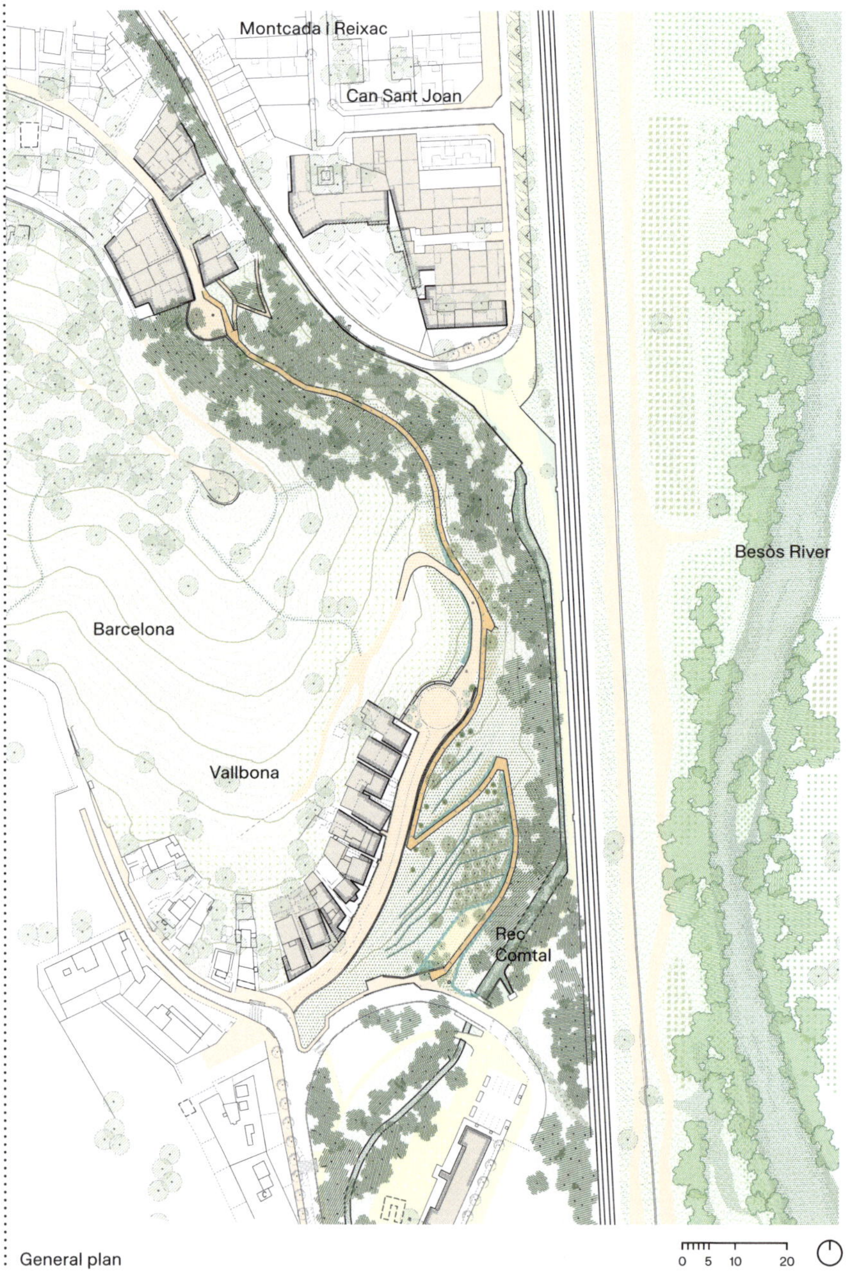

General plan

The new door-to-door sustainable mobility paradigm makes this path, which traces a line on the side of Vallbona Hill, a new, safer, healthier and more convenient connection between the two municipalities. The space is set to become a metropolitan link, a recreational route and even a natural lookout over the Besòs River.

Coexisting with nature

The pier-like footbridge on piles preserves the natural features of the site, facilitates the passage of wildlife and maintains the continuity of water along the natural runoff line. The two-meter-wide path avoids major earthworks while maintaining the topography of the place and respecting the existing pines.

Managing rainwater

In areas close to the urban fabric where the topography is less taxing, a dike-like path is created with a permeable gabion wall to retain and absorb surface runoff and stabilize the soil with its layers of broken slate.

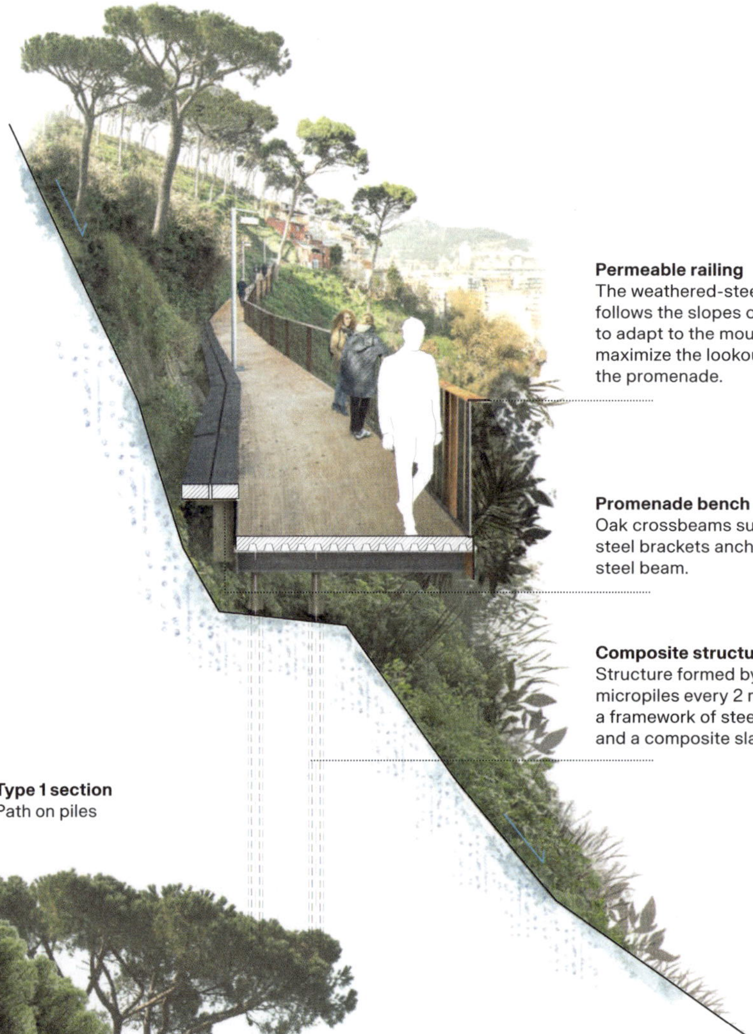

Permeable railing
The weathered-steel railing follows the slopes of the path to adapt to the mountain and maximize the lookout effect of the promenade.

Promenade bench
Oak crossbeams supported on steel brackets anchored to the steel beam.

Composite structure
Structure formed by a pair of micropiles every 2 meters and a framework of steel beams and a composite slab.

Type 1 section
Path on piles

Draining gabion
A draining dike that absorbs rainwater, reducing surface runoff on unstable soils, and fosters soil humidification along the path, which enhances the growth of flora.

Lookout walk
A concrete ribbon that adapts to the morphology of the site with an oxide-tinted concrete surface that, together with slate outcrops, is present along the route.

Type 2 section
Path adapted to the topography

Insect hotel at the start of the path

Client Consorci del Besòs / **Authors** Enric Batlle, Joan Roig, Ivan Sánchez / **Team** Clàudia Amías, Clara Terradellas, Marta Sanz, Laura Florentín, Elisabeth Torregrosa, Dolors Feu, Yago Cavaller / **Collaborators** SBS Simón i Blanco – Civil engineering / **Contractor** Copcisa / **Project area** 7,608 m²

Healthy Stairs

We can design buildings as an extension of public spaces, to host and promote social dynamics and be experienced and enjoyed like the open spaces we design for our cities. Buildings should promote healthy interior routes with access areas of sufficient size, highly visible bike parking spaces, attractive open stairways linking all the rooms, and landscaped terraces that interact with the outdoors.

The large open space designed in the existing Mercedes building enables the connection of all the areas of **the New Elisava School** by means of healthy staircases.

Efficient mobility interchange design

We need to promote door-to-door mobility (healthy mobility) by designing quality spaces that facilitate bike and scooter parking in new buildings. These spaces should be located next to the main entrance, at the same level.

Promoting sustainable mobility inside buildings

We need staircases that are easy to find and, if possible, have natural light and are spacious and ergonomic with vegetation and an elegant design; or stairways designed for atriums or courtyards that connect floors in a very intuitive way. We need staircases that encourage us to choose them over the elevator.

Encouraging healthy indoor routes and their connection with outdoor spaces

We need to incorporate interconnected courtyards and landscaped roofs in buildings by means of healthy routes and stairs that allow us to walk through the building from its entrance and mobility interchange points to all the spaces inside.

ESTADI JOHAN CRUYFF
FCB
FC BARCELONA
BARÇA, BARÇA
BARÇA!
MÉS QUE UN CLUB
FC BARCEL

Johan Cruyff Stadium in Sant Joan Despí, Barcelona (Spain, 2017-2019): an iconic, asymmetric image for a latest-generation stadium.

6,000 seats based on inclusive universal design criteria

458 m of complete perimeter walkway to access the stands

The new FC Barcelona Johan Cruyff Stadium forms part of the expansion of Joan Gamper Sports City, a project that we carried out in 2006. The stadium will host women's and youth subsidiary team matches with capacity for 6,000 spectators. The stadium, 137.5 meters by 110, adopts the old tribune idea to house most of the program and has a low U-shaped roof that closes the three remaining sides. In this way all the stands that surround the pitch are roofed right around the corners.

The main pedestrian entrance is on the lower level, forming a concourse where the press, VIP and staff accesses are located. Car-parking is concentrated on the upper level, in addition to the stadium entrances located at the four corners.

The exterior of the stadium is characterized by a neutral-colored base of precast concrete elements and perforated folded steel plates, contrasting with the red sheet metal that surrounds the upper levels. The dynamic façade, a lightweight second skin made of anodized aluminum discs with the diameter of a soccer ball, is located on the outside of the stands. The discs move slightly in the wind, like a sheet expressing what cannot be seen – the movement of the air –, a reminder that inside the stadium the ball is always moving. This skin provides solar protection while serving as a message display.

1

2

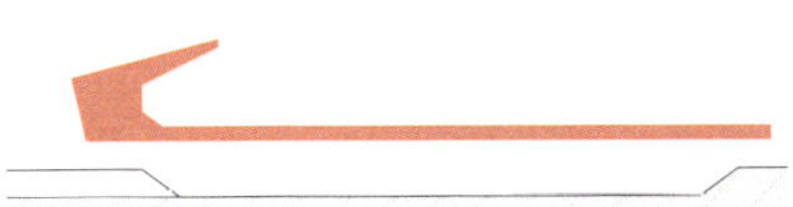

3

4

5

1 / Adaptation to the surroundings. The building follows the orthogonal orientation of the Sports City and adapts to the existing topography, allowing, along with the positioning of the entrances, for the removal of all architectural obstacles. The layout of the enclosure frees the stadium of barriers, corridors and dead ends.

2 / Asymmetrical shape. The desire to design a stadium in compliance with UEFA Category III standards required stands that incorporate television, lighting and VIP lounge facilities. The proposal integrates this substantial asymmetry in a single gesture, creating a uniform perception within the whole and constituting the iconic image of the project.

3 / Unity in a single gesture. Transition between two levels is achieved through a subtle gesture that transforms the lower-level coping into a grandstand railing, creating an interior belt around the pitch.

4 / Overall vision. The proposal seeks constant, overall vision throughout the facility. This makes it easy for visitors to find their way around. It facilitates easy movement and an intuitive understanding of how the building functions, with coherent circulation and clear signposting. The stands are arranged on two levels for the best views and maximum closeness between players and spectators.

5 / Greek theatre. Following the classical standards of Greek theaters, spectators always access their seats from the upper level, through an ambulatory that surrounds the stadium. This gives spectators a general view of the stadium on their arrival and facilitates their understanding of the space.

Spectators reach their seats from the parking lot via a continuous accessible ambulatory that surrounds the stadium and leads to all the sections in the main stands. Connection between the outside and the interior is continuous, easy and direct, thanks to the numerous entrances provided all along the perimeter. The fences were designed to be highly transparent to blur the public-private boundary and facilitate communication.

Cross section

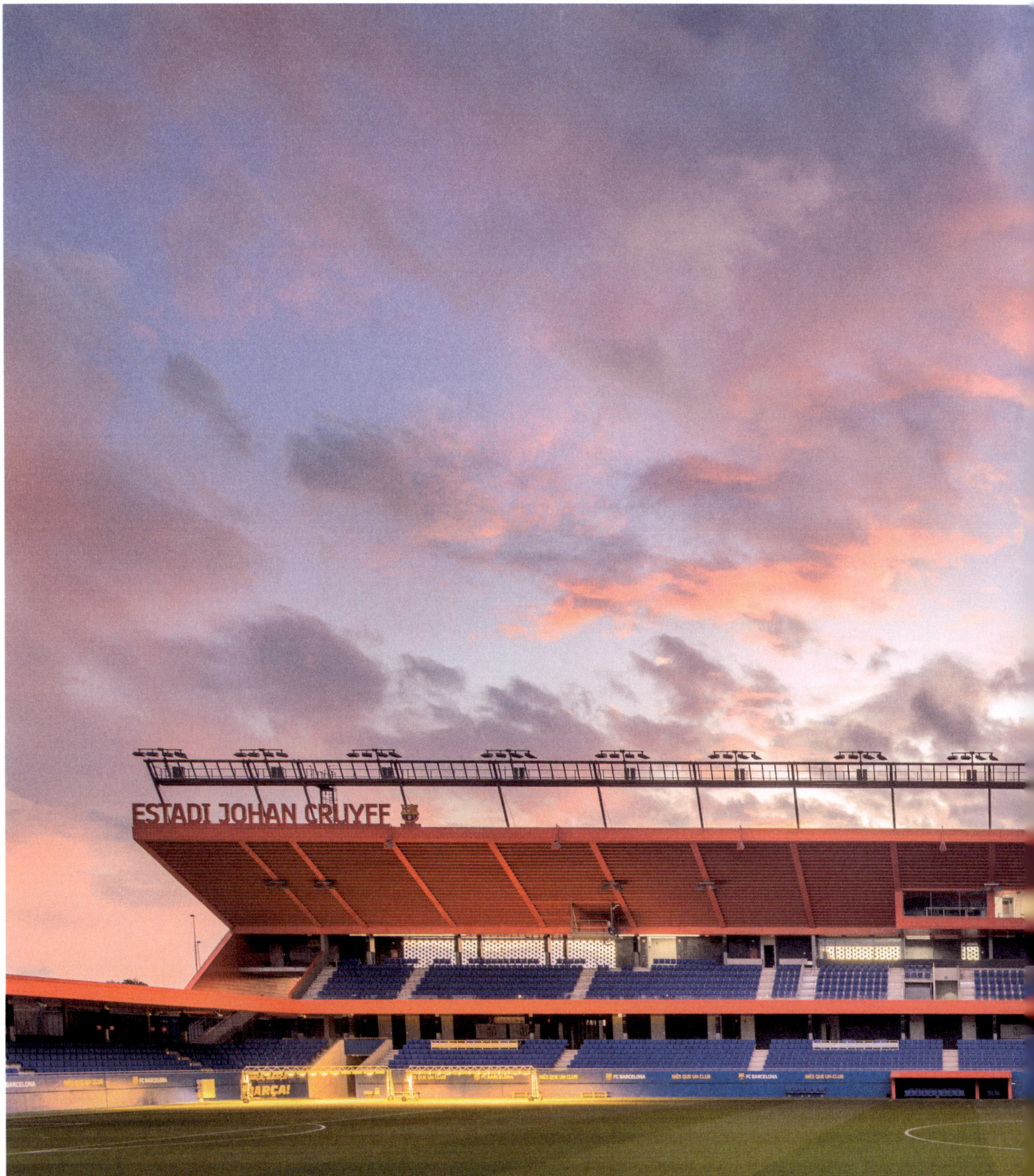

Johan Cruyff Stadium is a project that respects the diversity of all its users: healthy, safe, honest, inclusive, smart and sustainable. These values won it the 2019 Good Practices Award of the Design for All Foundation.

Client FC Barcelona / **Authors** Enric Batlle, Joan Roig, Albert Gil / **Team** David Sesé, Francisco Javier Luna, Francesc Puig, Marta Peinado, Joan Batlle, Mercè Lorente, Carlos Griell, Nuria Hernández, Yago Cavaller, Maria Duran / **Collaborators** BIS structures – Structures Engineering, PGI Engineering – Facilities Engineering, ARCO – Façade Consulting / **Contractors** UTE Estadi FCB JC, Serom / **Gross floor area** 7,382 m², 18,687 m² (urbanization and parking)

Building

Living on the Border Between the City and the Country

The city and the countryside have traditionally had a very poor relationship. The city gobbles up all the territory it needs, imposing its laws and eliminating all the systems previously established there. We must advocate for a paradigm change in the model that will lead us to food governance in cities.

Urban agriculture is the most economical maintenance system for open spaces. It is the image of landscapes we long for and can become an urban philosophy, as in the case of the Incredible Edible movement, which has transformed all the open spaces – from allotments to gardens, from squares to schools – in the English town of Todmorden into productive spaces for local food production.

Productive landscapes reclaim biodiversity, clean the air, manage water, contribute to the generation of new energies, can lead to the creation of new materials and produce an abundant diversity of foods, from urban gardens to community agriculture.

Productivity

Food Governance

We need to place value on the productivity of the urban infrastructure in our cities to guarantee the preservation of their extension and quality, seeking to construct a New Green Economy that promotes the creation of new materials or energies, maintenance of the water cycle and local food production with the aim of moving closer to food governance. By creating sufficiently sized open spaces as a system linked to watercourses that preserve the agricultural structure, we can transform all the urban interstices in our large cities into productive infrastructure.

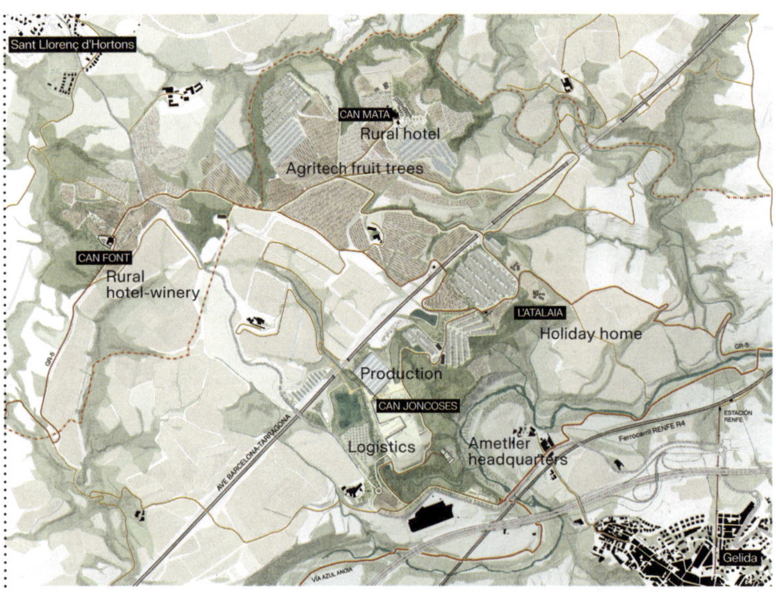

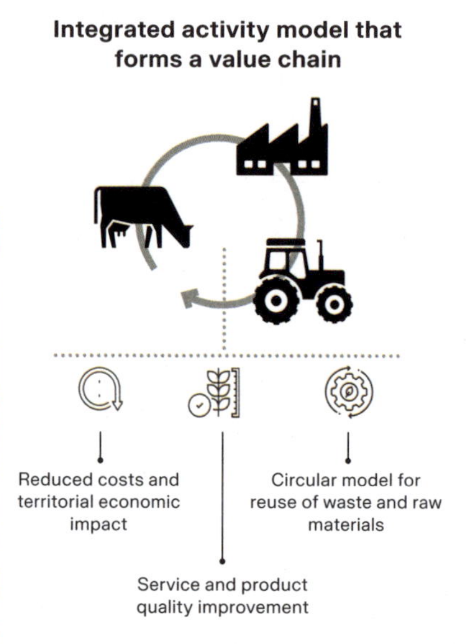

Agroparc, located at the entrance to the metropolitan area of Barcelona and owned by the company Ametller, seeks to implement in a single territory the entire agrifood chain: from primary agricultural activity to transformation of the product to its direct distribution. Unmistakably agro-industrial in nature, the project integrates high-tech agriculture to manage outside factors and traditional agriculture, which is the basis for the dissemination of the environmental values of the territory where it is located, Penedès. Sensitivity to the preservation of the unique landscape of the region, based on sustainability, innovation and circularity, rounds out a project that seeks to be exemplary in the integration of the entire agrifood sector cycle.

Creating open spaces of sufficient size that are close to citizens

Arranging open space as a coherent system connected to the geography makes it a central metropolitan space and, therefore, close to citizens, who can use it not only for leisure but also for production.

Maintaining agricultural structures and historical irrigation systems

The transformed preservation of former agricultural areas as open urban spaces also makes it possible to maintain their subdivision and irrigation structure, taking advantage of them to continue contributing to food production while also serving as a motivation for urban planning.

Rethinking the public space management model

A maintenance model focused on gardening and cleaning conducted by the public administration can give way to conservation with shared responsibility with citizens who, through cultivation and use, participate in the fruits of their labor in exchange for reduced maintenance needs.

Development of the Finestrelles Sector in Esplugues de Llobregat, Barcelona (Spain, 2004-2022): a new green gateway connecting the city and Collserola Natural Park.

+15,000 m² of alfalfa meadow

20 ha of green area with native species connected to Collserola Park

3 km of undergrounded high-voltage lines that serve Barcelona

Current planning, 2004
Uses

- Open space
- Developable land
- Road network

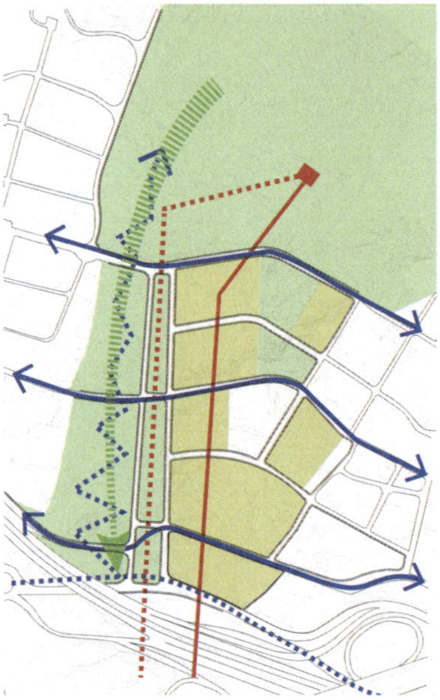

Development planning. 2004
Action strategies. Uses

- The central meadow, a large traffic circle designed with the landscape in mind, is used to underground the power line.
- Developable land
- Connection between the city and Collserola Park
- Transversal connectivity between existing road networks
- Soft connectivity between the city and Collserola Park
- Existing high-voltage line
- Undergrounding of the high-voltage power line

Green space system

In Finestrelles we were carrying out the urban planning of the sector and the development of all the public spaces in the area. The transformation of the urban vacuum left between the Finestrelles, Ciutat Diagonal and Can Vidalet neighborhoods, located at the entrance to Barcelona via Avinguda Diagonal, entailed burying the power lines that crossed it. From the start, it presented an opportunity to transform existing planning, which had envisioned the development from the point of view of automobiles with traffic circles for road crossings. The proposal involves an arrangement that links the city and Collserola Park more closely. The urban development widens its streets and is structured around a large park which, like a green wedge, extends the nearby natural spaces towards the interior of the city. The park serves as a public access corridor from the city center to the network of paths that lead into Collserola Natural Park. The central meadow is arranged as

a large traffic circle designed with the landscape in mind and through which the undergrounded high voltage line runs.
We decided to use native vegetation in the new park. We wanted the country to enter the city and to reinforce this idea, we decided to plant a large natural meadow with a mix of seeds including alfalfa in the center of the urban development. The choice of plants proved to be correct. The site has held up quite well, despite not being watered or cared for by anyone for more than seven years.
We could hardly believe it when one day, we saw how, occasionally, a flock of sheep would come down from the nearby hills to graze in the new urban meadow. After the requisite monitoring, we have confirmed that shepherds are still active on our urban periphery and that pastures are few and far between in this uncontrolled magma of unbridled infrastructure, large facilities and unfinished urban areas we call a city. Cayetano, our shepherd, also was pleasantly surprised the day he discovered the succulent fare that the city had to offer his flock for free.
Perhaps the real-estate crisis provided us with a period of reflection that we used to learn new forms of behavior. The image of a flock of sheep is an allegory of a city that could exist in harmony with its surroundings, a small sign that perhaps we can find new models of co-existence between the city and nature.

General plan

Section

Ridges with aromatic plants to retain water

More than ten years later, the area is under development, and we have had the opportunity to participate in some of the projects. The unique arrangement of the public space serves as a point of reference for its architecture.

A flock grazes in a new public space in metropolitan Barcelona
We could think this is an everyday occurrence, similar to what we might observe on our urban peripheries, places already touched by urban expansion but where this development does not prevent natural exteriors from infiltrating the city. We could also interpret the image as a symbol of yet another consequence of the economic crisis of recent years. The place was undergoing urbanization with the aim of introducing a business district and residential community, but when the real-estate bubble finally burst, the project was abandoned before a single building was constructed. The streets and public spaces in the area were nearly finished and operational, but because of this abandonment, they were neglected and left unmaintained for many years.

The large central meadow is, simultaneously, a productive space that feeds one of the last flocks of sheep in Barcelona, a large roundabout that distributes traffic in the area, a protective space set aside for the undergrounding of power lines and a green corridor that links the city with Collserola Natural Park. Municipal regulations did not provide for the possibility of a flock using a public space. At first, the local police prohibited the flock from entering the meadow, assuming this activity undermined the use of the site as a public space. As a result, we organized a meeting between Cayetano –our shepherd– and the local authorities to try and find a regulatory mode of coexistence between the two uses. It is what the general population wanted, and the news was widely reported in the media.

Meeting between Cayetano, Enric Batlle and the Mayor of Esplugues de Llobregat, Pilar Díaz.

Discover Les Rondes Beltway hackathon. Barcelona Regional, October 20, 2018.
A group of students from different European architecture schools met with shepherds and their sheep to illustrate how people, bicycles and animals can share the same urban space in a respectful and complementary way. It served to show that urban spaces can be productive, biodiverse and connected.

Client Compensation Board of the Sector / **Authors** Enric Batlle, Joan Roig / **Team** Abel Porcar, Iván Sánchez, Clàudia Amías / **Project area** 395,200 m², 234,800 m² (Gross floor area)

City and Territory

Zero Km

We need to design productive gardens in our public spaces that generate local food products for citizens. Prioritizing the use of local materials and solutions benefits the project itself by reinforcing its identity, uniqueness and integration, in addition to contributing to the local economy. Looking to the agricultural traditions of our culture, we should seek out the tools we need to create the best public spaces possible.

The projects developed for **Green Riyadh 2030** incorporate large extensions of public space located between the road junctions of major infrastructures, with new productive plantings, trees and other plant species, which will serve as a nursery for the landscaping of streets and other open spaces in the city. The borders of large urban avenues and roads are also designed based on successive plantings, which, in time, will densify arboreal vegetation in open spaces.

Thinking of the landscape as a useful and productive space

We need to reclaim public space for use by people while embracing production strategies that give it management self-sufficiency. Agricultural dynamics such as kitchen gardens and nurseries are compatible as landscaped areas and provide a usable surplus that impacts the city itself, giving character to the landscape beyond the merely aesthetic.

Incorporating agricultural dynamics in green spaces

Sustainable agricultural techniques contribute to the environmental improvement of the surrounding areas. Introducing these techniques into urban open spaces entails encouraging crop rotation based on complementary species and the resulting variation of the landscape throughout the year. This dynamic of change transforms green spaces all year round and creates greater adaptability according to social needs.

Activating the trophic chain

Promoting plant variety based on alternative and complementary planting involves a seasonal chain linked to the flowering season and harvest, including the transplanting of species in the case of plant nurseries, to reactivate the biological dynamics of the place. Carbon and organic agriculture are good examples for increasing biodiversity, both of pollinating insects and indigenous microorganisms in the soil, and even the production of biofertilizers and mineral-based compounds.

Recovery and Improvement of Boca de la Mina Historic Promenade and Creation of the New Aromas Park and Agricultural Garden in Reus, Tarragona (Spain, 2018-2021): a historical promenade as a healthy space in the city that places value on its century-old identity and promotes its agricultural memory and contact with the Camp de Tarragona landscape.

A series of **11** urban gardens where nearby schools can teach the values of agriculture

A typical plantation of **78** fruit trees in Camp de Tarragona

Farmland with **23** different species of aromatic plants

The Boca de la Mina promenade originates from a farm road that connected the urban center to the surrounding land and groundwater mine after which it is named. Since the end of the 19th century, it has also increasingly been used as a place for strolling and leisure by the local population, lodging itself in the city's collective imagination as an emblematic space with immense natural, heritage and sentimental value.

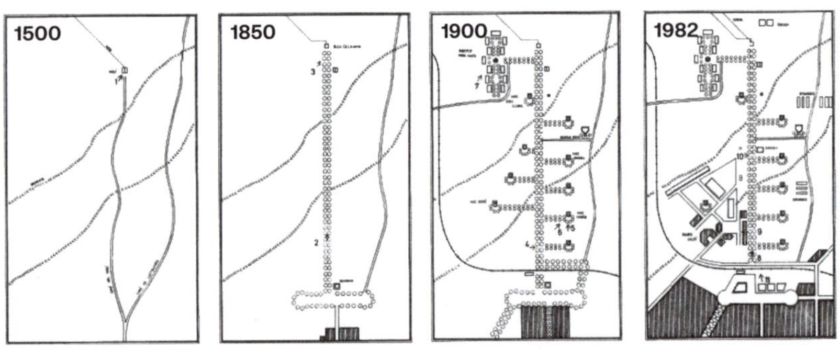

Historical evolution of the promenade

General plan. The promenade as territorial connector

The proposed improvement and refurbishment of the promenade is defined by the work on its standard section. Its transverse structure is modified to direct drainage to lateral landscaped areas through the creation of a system of green ditches and water infiltration areas.

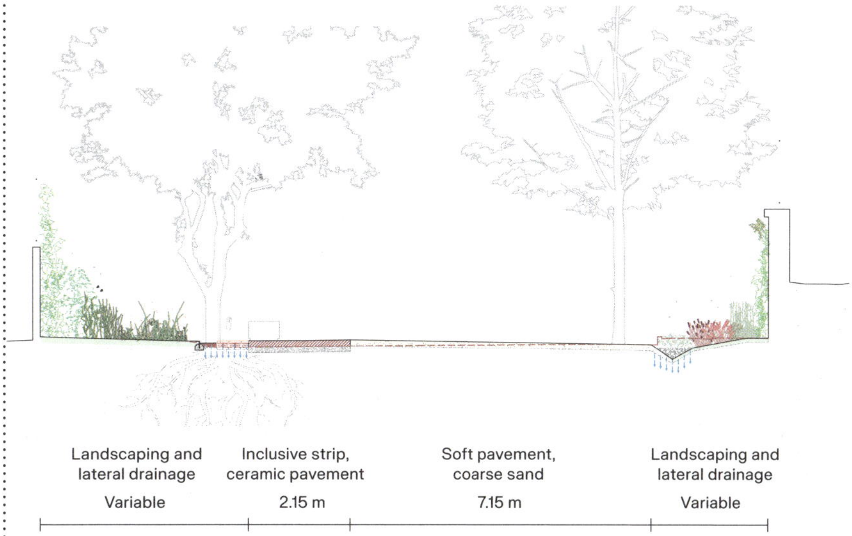

Type walkway section

The mass of plane trees that provide shade and personality to the path is preserved and complemented. The paved brick border and new lighting dramatically improve the accessibility and inclusivity of the route without significantly altering its original earthy character. Brick, in different formats and sizes, is the basic material used for the paving, furniture and other elements along the promenade. This ceramic material is a perfect fit with the variety of existing elements of *Modernista* design such as the helicoidal chimney at Mas del Molí, the perimeter accesses and walls of several plots of land, and the Pere Mata Institute.

A series of new spaces complements the promenade along its route:

Aromas Park. A new place for rest and meditation that links the promenade to Molí ravine. A system of pedestrian routes that provide a sensory experience through the cultivation of aromatic plants arranged by color.

Agricultural park. A public space structured as a farming plot for productive, educational and recreational use. An orchard of fruit trees and a vegetable garden where we can learn about the agricultural values of Camp de Tarragona.

Bassa del Bacallà belvedere.
A small metal structure offers a new vista of the pond. It takes its name from its shape, reminiscent of dried cod. It was one of the places where young people used to swim on Midsummer's Eve. At the beginning of the 20th century, this pond represented an important engineering feat, collecting water from Riudecanyes reservoir to meet the city's needs.

Client Reus City Council / **Authors** Enric Batlle, Joan Roig, Iván Sánchez, Mario Súñer / **Team** Roger Mermi, Gabriel Romero, Clara Terradellas, Marta Sanz, Esther Díaz, Bárbara Freire, Lina Florez, Marta Ines, Laura Luque, Gemma Ramon, Yago Cavaller / **Collaborators** SBS Simón i Blanco – Civil Engineering / **Contractor** Valoriza Medioambiente / **Project area** 42,000 m²

Productive Buildings

Each urban rooftop and terrace is an opportunity to attain a productive city. We can commit to recovering the idea of the *masia*, the traditional farmstead, constructing productive buildings that generate the bulk of what its users consume: from local foods to electricity; from the air we breathe to the rainwater we can collect.

At **the Headquarters of GB Foods**, part of the Agrolimen group, landscaped terraces were designed to be used as workspaces. In the part corresponding to GB foods, aromatic plants were chosen for the different kitchens throughout the building while in the Affinity part – pet food – plants were chosen that encourage insect reproduction.

Promoting productivity in buildings

We can understand the productivity of a building as its capacity to meet all its users' needs. In this way, the ideal productive building should produce food, oxygen and electricity, and store water without depending on outside sources.

Connecting buildings to productive environments

The traditional Catalan construction of the *masia*, the farmstead, consisted of surroundings that managed available water, cultivated the necessary food in its own vegetable garden and farm, and used animal body heat and the kitchen fire to heat the house.

Rooftop farm

The rooftop of a building can become a productive space that provides users with healthy local food in a sustainable and environmentally friendly way, contributing to a greener and cleaner city in contact with nature. From a small terrace of a residential building to the large rooftops of office buildings or shopping centers, buildings represent an opportunity to create these types of spaces, bringing food production centers closer to consumers.

Expansion of Roche Diagnostics Headquarters in Sant Cugat del Vallès, Barcelona (Spain, 2018-2022): an umbraculum building in a renaturalized environment.

+4,000 m² of productive green roofs

50% and **38%** reduction in concrete and steel, respectively, using a structural system of Holedeck lightweight beams

100% natural ventilation of 3 basement parking floors

The expansion project is located on a 38,000 m² plot of land in Can Sant Joan Business Park where the Roche headquarters already stood. The proposal includes the construction of two buildings and a landscape development with a large 3,500 m² green area that serves as a meeting point for all the buildings on the plot, which will become a large central square on the Roche campus.

The first building, a parking area to meet the needs of the building complex, is circumscribed by two ventilation courts and an additional courtyard in a central position. The result is a highly energy-efficient building and the simultaneous reduction in the need to excavate. It is the only underground parking area with natural ventilation in Catalonia. The second building has a rectangular floor plan and is intended for the new offices. It is situated parallel to one of the borders of the plot along a south-north direction, creating an ecological corridor throughout the center of the structure that joins the natural spaces of the two adjacent streams.

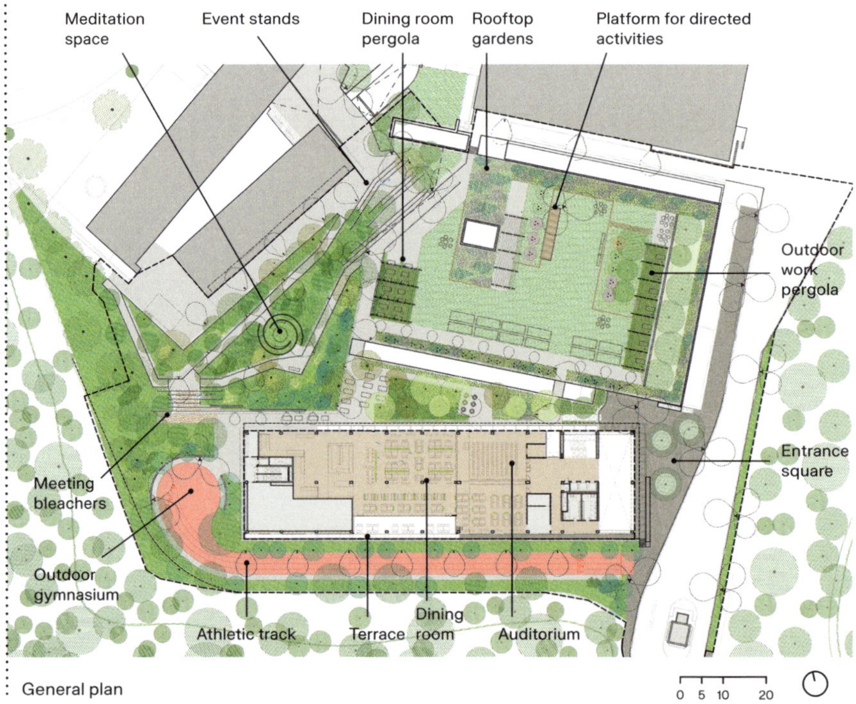

General plan

Starting with the design phase, different strategies are implemented to make Roche's expansion a productive and environmentally responsible complex: rooftop gardens, own energy production, use and reuse of rainwater and gray water, and passive strategies to reduce energy demand.

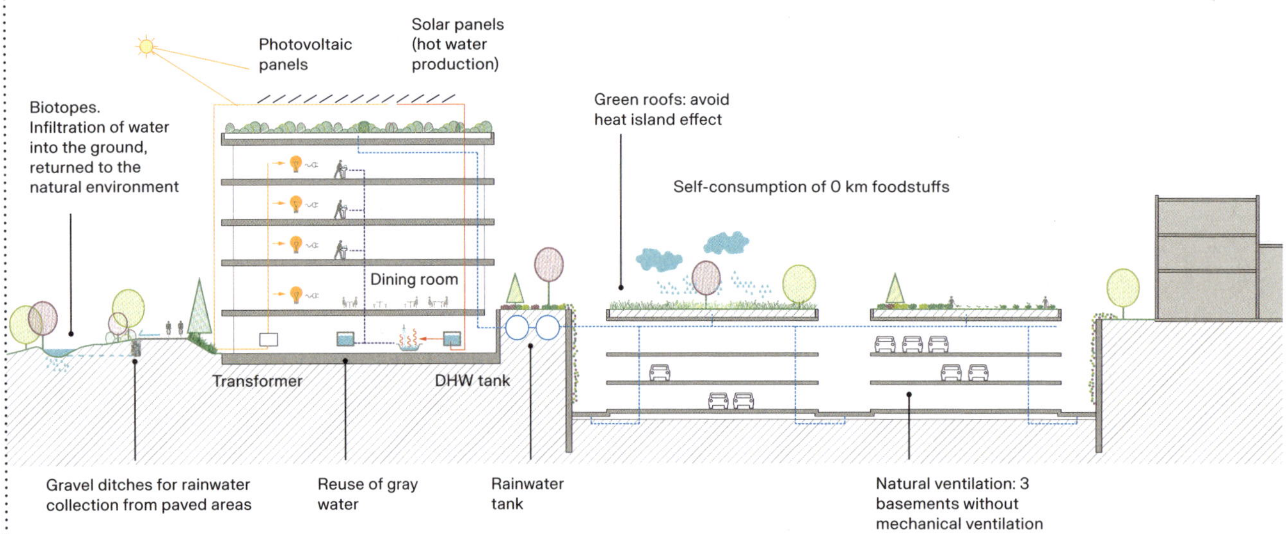

Section

The central square on the Roche campus houses a large green area on the roof of the garage floors, with plant pergolas that provide shade for outdoor work spaces, a large esplanade with a stage for organized activities, and, taking advantage of the topographic difference, stands for ceremonies and presentations. The landscaping design connects the different buildings that make up the campus and interrelates with the nearby green corridors.

The roof is designed as a green extension that will provide thermal inertia to the interior of the building.

Access to the office building is located at one end of the semi-basement level through a double-height hall. On the ground floor, at the same level as the central square, are more public-use facilities: the canteen and auditorium. The offices occupy the upper floors and are arranged around a central axis. The building façade consists of a double skin inspired by traditional slatted facings that adapt to each orientation to provide solar protection. The interior skin is a modular glazed façade system, while the exterior skin is made up of a steel frame that supports the wooden slats that have been heat-treated to make them stable and durable.

The structure was built using the innovative Holedeck system, which uses site-cast reinforced concrete beams that are lightened by inserting holes along their entire length. These openings house installations inside the structure with the resulting decrease in required material compared to traditional concrete systems, thereby streamlining the use of materials to make the system more sustainable. Both the structure and installations are visible.
The interiors are based on constructive honesty and the use of traditional local materials to combine technology and sustainability with the domestic and local sphere. The use of vegetation is prevalent in interior spaces to promote users' well-being.

Client Roche Diagnostics / **Authors** Enric Batlle, Joan Roig, Albert Gil / **Team** José Sanz, Josep Batlle, Anna Porta, Joan Batlle, Marina Santos, Mercé Lorente, Toni Duran, Diana Calicó, Carolina Valero, Gemma Ramón, David Sobrino / **Collaborators** BIS structures – Structural engineering, PGI Engineering – Facilities engineering, Envolventes Arquitectónicas ENAR – Façade consultancy, Betarq Group – Building engineering / **Contractor** Elecnor – Offices and urban development, Carbonell Figueras – Car park / **Gross floor area** 10,322 m² (offices), 12,133 m² (parking), 9,500 m² (urbanization)

Regenerating the City and the Territory

If we rethink the future of the city by proposing systems that permanently reverse the current models of unlimited extensive territorial consumption and polluting mobility based on private vehicles, we discover two areas of action with many possibilities: urban regeneration and the creation of boundaries between the existing urban fabric and the green infrastructure that we want to preserve.

Most of our cities require urban regeneration processes to improve their conditions or to start a new useful life that offers fresh uses and avoids consuming more territory: from the obsolete industrial areas that need to be reprogrammed to the slums that need to be rehabilitated, from the aggressive infrastructures that we need to redesign to the new mobility interchanges we have to promote, from the urban centers that require traffic calming and renaturalization to the sectors that can still be densified.

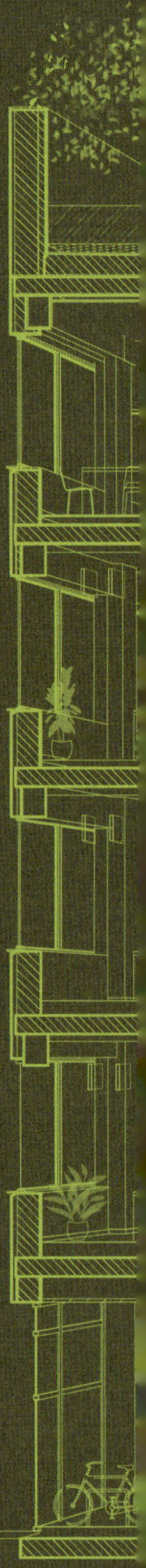

Recycling

Urban Recycling **Revaluing Place** **Rehabilitating vs. Demolishing**

Urban Recycling

Improving urban centers, reinforcing declining urban fabrics and reprogramming obsolete infrastructure are operations with numerous beneficial consequences: at the environmental level, they reduce the consumption of resources and land consumption for city expansion; at the economic level, they make use of existing resources and recompense past investments; at the social level, they strengthen neighborhood roots and preserve cultural heritage.

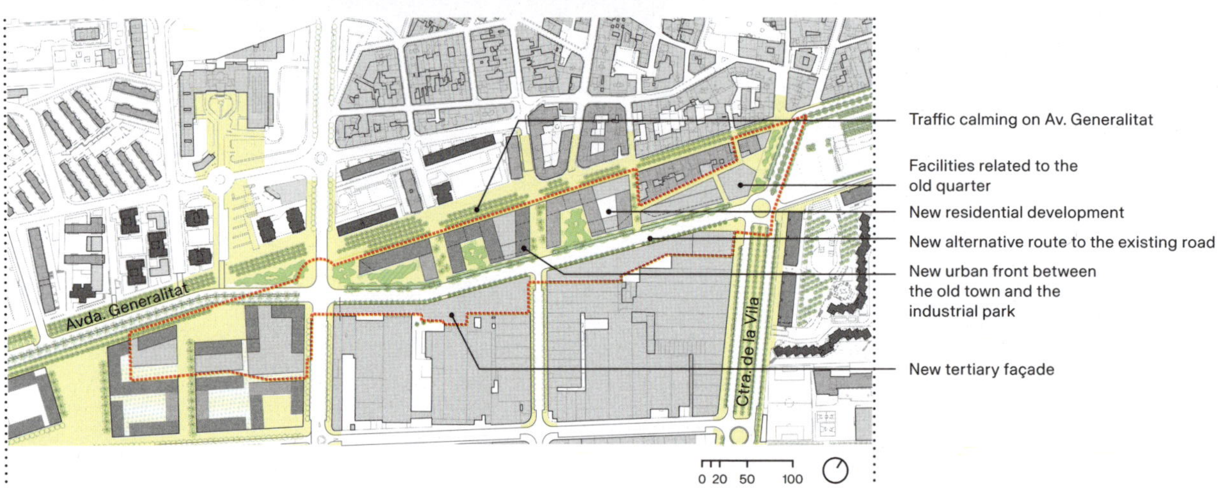

The proposal for **the Surroundings of Avinguda Generalitat in Viladecans** involves the development of what is currently an abrupt border between the industrial area and the city center by creating a juncture made up of new tertiary and residential activities, open spaces and the diversion of vehicular traffic, gaining civic space on the old C-245 highway. One of the first sites of growth in Viladecans, it was later absorbed by the needs of an industrial park. Now it is undergoing a third transformation to resuscitate its civic centrality.

Avoiding conventional urban transformations that obliterate all pre-existing elements

Before knocking things down, it is important to assess the possibility of maintaining and rehabilitating: the layout of streets, sewers, planted trees, historic buildings, infrastructures, etc. All the existing elements in our cities have economic, social and even sentimental value that we should not ignore.

Strengthening immediate, quality economic solutions

The use of existing elements in smaller scale interventions adapted to urgent times allows us to further the enjoyment of new spaces while avoiding periods of decadence or impasse, maintaining levels of urban quality comparable with those brought about by classic urban transformations yet with far more moderate transformation costs and times.

Initiating the renaturalization of the city while preserving previous urban structures

Like large metropolitan watercourses, the existing city has wildlife and flora habitats that must be used as a starting point in any urban transformation: former crop fields, trees from urban or agricultural plantations, private and public gardens, irrigation channels, and low-lying and wet land.

-ROC BLACKBLOCK-2019
Volgueren enterrar-nos
no sabien que érem
llavor!

Provisional Urbanization of the Industrial Area of Can Batlló in Barcelona (Spain, 2013-2016): an alternative sector development model adapted to the unexpected economic circumstances created by the financial crisis, leading to the reactivation and regeneration of the industrial premises and benefitting from the transformational momentum of neighborhood groups and movements.

15,000 m² of tactical urban planning, making a 6 ha site operational

+20 social initiatives introduced in all industrial premises and outdoor spaces provided

6,000 m² of new green areas, with 20 new species and site permeabilization

The Can Batlló industrial site covers nine hectares in the heart of the Bordeta neighborhood. Along with Vapor Vell and Vapor Nou, it was one of the first economic motors in Sants at the end of the 19th century. The textile crisis led to the initial transformation of the site into a "commerce city" with more than two hundred small businesses that allowed it to remain operational until the late nineties.
However, with the advent of the new century, its activity decreased dramatically. At this time, a second transformation process, implemented at the height of the real-estate boom, anticipate a significant urban development with a new residential fabric but also a substantial number of facilities and a large park. In 2008, the financial and real-estate crisis suddenly, and indefinitely, brought this transformation to a halt. After three long years of social demands to restart the change, a neighborhood platform entered the site on June 11, 2011, at 11.00 a.m., with the aim of organizing a series of recreational and protest actions and beginning the process of reclaiming the sector for the neighborhood, even if it meant using its own resources.

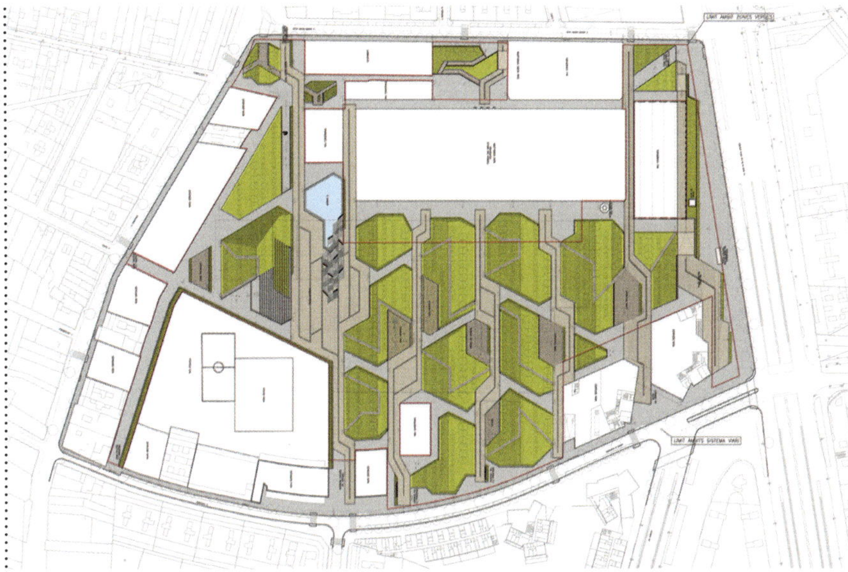

Urbanization project approved in 2000 and stopped by the crisis. General plan

Provisional urbanization project approved in 2015. General plan

In 2000, we proposed a new urban park built from scratch with preexisting assets. The provisional project, executed with a minimum budget, recycles what is there and places value on the industrial units on the site, leaving interstitial spaces for neighborhood activities. Perhaps all that remains is to understand which actions of this kind could be permanent, and the fact that the future can be built without having to destroy the past every time we want something new.

Existing

The proposal is set in the framework of the urgent social context in which conventional urban models are not effective. Instead, tactical urban development actions are proposed to provide a clear, quick response to social demands. A system of routes based on the preexisting network of streets, where all accessibility and safety issues are addressed, is proposed. The industrial premises are equipped with the necessary service networks, and plots of land are created where new uses can be developed. In addition, a major effort is made to move forward with the renaturalization of the inhospitable industrial environment.

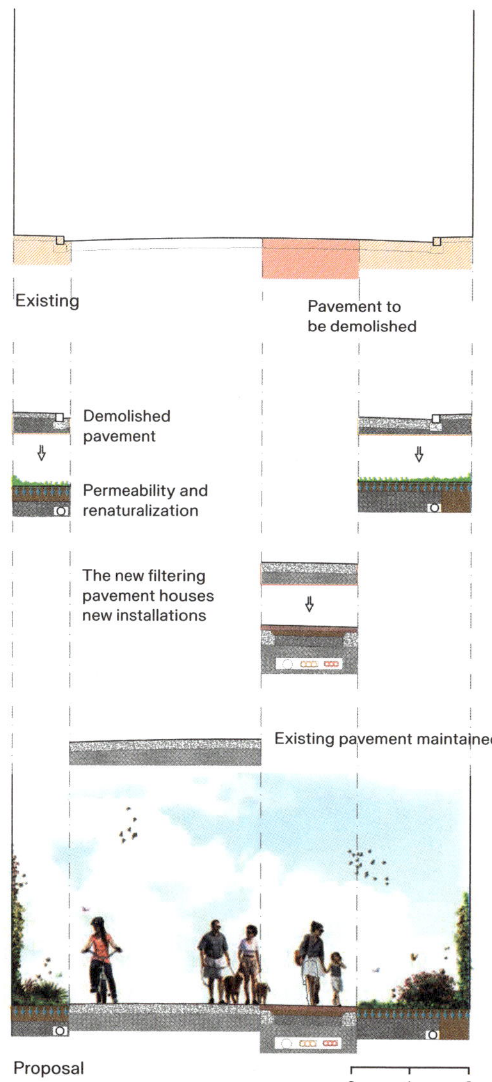

Carrer 11 de Juny before and after construction work. Section

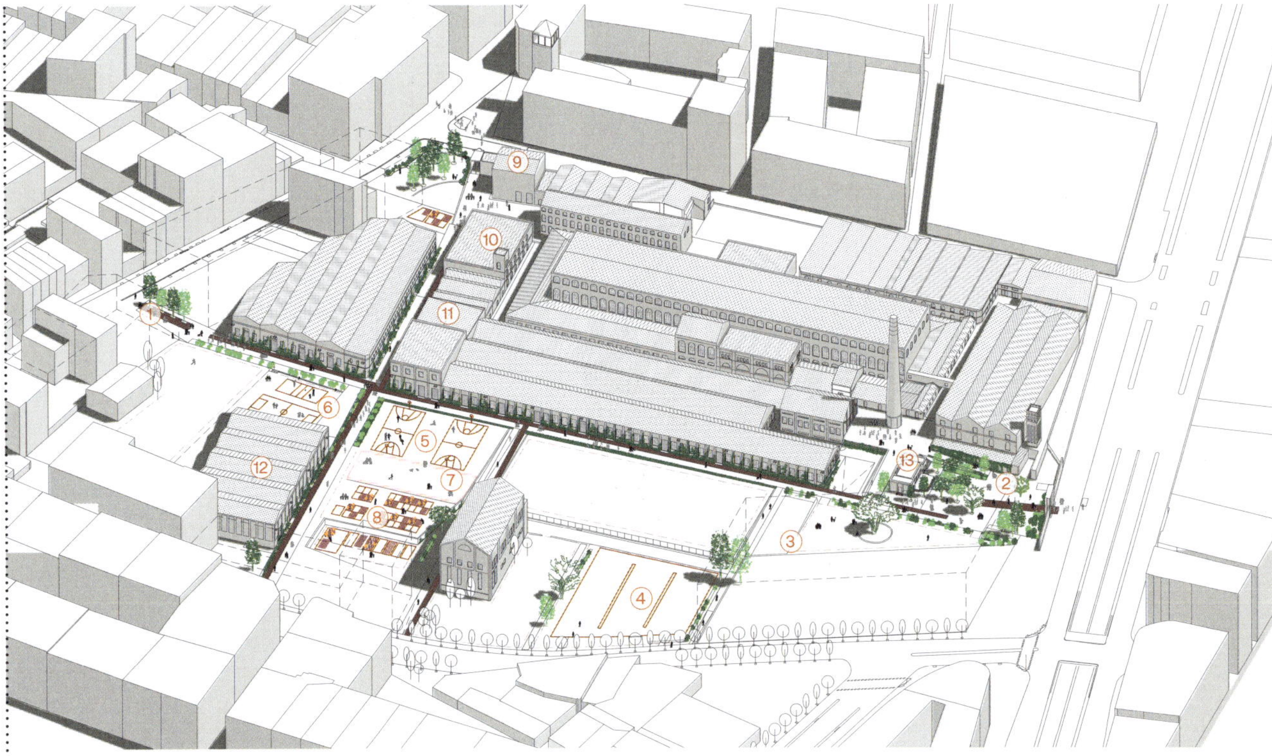

The Can Batlló industrial site

1/ Square leading to Carrer Constitució
2/ Square leading to Gran Via
3/ La Olivera Garden
4/ Car park
5/ Sports courts (basketball)
6/ Sports courts (soccer and volleyball)
7/ Play area for dogs
8/ Urban vegetable plots
9/ Old chapel - ecological cooperative and exhibition space
10/ Auditorium, bar and popular library
11/ Block 11 - Arts Can Batlló, multipurpose spaces and workshops, printing, carpentry, etc.
12/ Block 9 - mobility center, family space, performing arts space, etc.
13/ Canteen

The newly developed infrastructure has led to the reactivation of various industrial units by numerous citizen initiatives that have introduced such projects as the library, auditorium, bar and canteen, in addition to different workshops, a family playroom and a performing arts center. A good part of the outdoor spaces has been set aside for recreational uses such as sports fields, a dog run and urban vegetable gardens.

The new protagonists of the Can Batlló site

The old Can Batlló industrial site has become a nucleus of social and leisure activity within the Sants-Montjuïc district. The walls and fences have come down and new routes have improved overall connectivity in the Bordeta neighborhood. Its architectural heritage is under guardianship once again thanks to its reactivation during this transitional time.

Client BIMSA – Barcelona d'Infraestructures Municipals / **Authors** Enric Batlle, Joan Roig, Iván Sánchez / **Team** Mario Súñer, Dolors Feu / **Contractor** Construcciones y Servicios Faus / **Project area** 18,700 m²

Revaluing Place

Preexisting elements of place are reassessed to preserve its identity, achieve better landscape integration within the surroundings and optimize the resources used. It entails respecting the inherent character and values of the place, with the understanding that, when designing a landscape, we never start from scratch.

The Banks of the Llobregat are covered with reeds that are systematically eliminated because they are considered an invasive plant. Meanwhile, new paths and lookouts are being built that require the use of non-local materials. In the Pallejà section of the project, we are experimenting with the reuse of reeds to form boundaries and small earth-retaining structures.

Strengthening the inherent values of the place

Preexisting elements are reassessed according to numerous criteria: functional, aesthetic, cultural and historical. Actions are proposed that place value on significant elements and avoid what is improper or secondary. Close attention is paid to the energy costs of each decision to avoid generating unnecessary waste and significantly reduce the carbon footprint of the action, promoting a form of design based on a closed, environmentally-friendly circular process.

Consolidating the identity of the place

The values of the place are identified as a departure point, recognizing important factors to create an open space with its own personality. On-site archeological findings are incorporated into the project to preserve its essence while the design integrates the culture and history of the place to create site-specific work.

Contributing to the evolution of the territory

A preexisting foundation will always be more resilient than a new design, as it has transformed and evolved over time according to the dynamics of the place. Respecting existing vegetation, even the original topography and water dynamics, becomes an opportunity to design a space with an identity that is also adaptable to the passage of time.

Plaça del Cobert de la Màquina de Batre in Sant Fruitós de Bages, Barcelona (Spain, 2016-2018): a new multifunctional public space made up of gardened terraces that complement the large roof that used to house a threshing machine and commemorate the town's agricultural past.

360 m of agricultural terraces in the form of an amphitheater

100% rainwater infiltration

The unique roof dates back to the 1950s, when the municipal Chamber of Agriculture decided to build this reinforced brick structure to house the threshing machine that was used to separate the seed from the stalks and husks of the harvests of local farmers. In the second half of the 20th century, its use fell into gradual decline due to the industrialization of the region, crop changes

and the emergence of new technologies. At the same time, the structure was absorbed in piecemeal fashion by a section of the town's expansion that failed to make provision for open space around it.

The proposed project solves the problem of contact between the roof and the town by creating a new public space to serve as a central square for leisure activity for local inhabitants and large-scale events. Its design and development, inspired by rural terraces and dry-stone walls, sets out to highlight the value of this structure by recalling its agricultural past. Heritage and historical memory merge in a project for the town's future.

Main project strategies

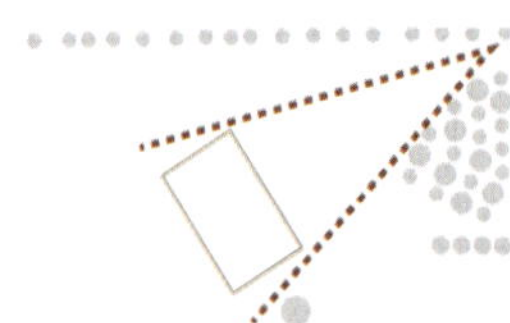

The roof as protagonist
The arrangement of all the elements of the new square revolves around this unique architectural element, both protagonist and focal point.

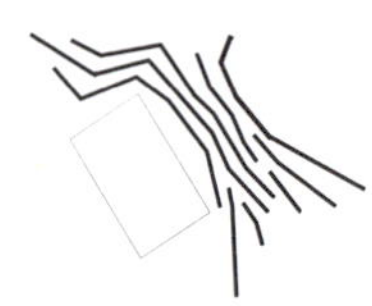

Agricultural terraces make up the new public space
A series of terraces shape this multipurpose space where visitors can observe, rest, play or walk. The dry construction of gabions reflects the traditional agricultural landscape of terraces and dry-stone walls in the area.

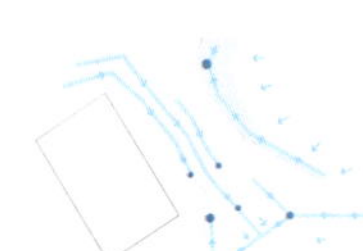

Water management as a resource
The new landscaped terraces define a drainage system based on infiltration and the use of rainwater.

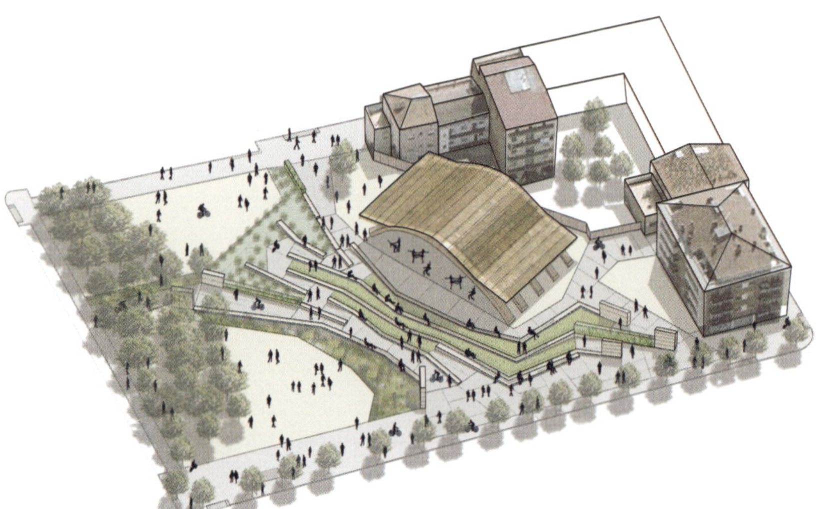

Axonometric view of the project

Client Sant Fruitós de Bages Town Council / **Authors** Enric Batlle, Joan Roig, Iván Sánchez / **Team** Mario Súñer, Simon Gabillard, Lidia Carrillo, Dolors Feu / **Contractor** Constructora Calaf / **Project area** 4,500 m²

Rehabilitating *vs.* Demolishing

Rehabilitation is a fundamental principle of sustainability as it enables extending and improving the useful life of a building, making up in half the time the environmental impact caused by its original construction. Refurbishing obsolete and inefficient buildings, instead of razing them and building anew, significantly reduces the environmental impact, the resulting energy consumption and the generation of waste.

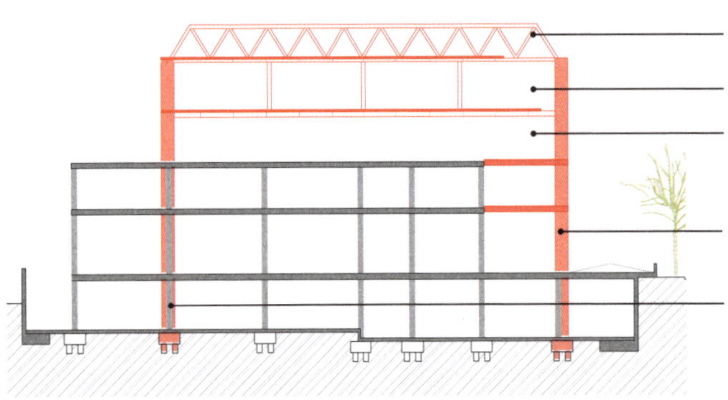

The design for an **Office Building in the Marina del Port District in Barcelona** transforms an existing building with little architectural value designed to house cars into a building conceived with people in mind. The new structure was designed to meet the expansion needs of two new levels while minimizing the impact on the existing structure. To do so, 32-meter-long metal trusses have been positioned on pillars that make up the new building façade.

Valuing what is already built

It has been common practice to value and rehabilitate only preexisting architecture of heritage interest, and to demolish and rebuild the rest. This approach ignores the fact that all architectural structures have had an environmental impact (carbon footprint) in their construction. Rehabilitation is an attitude that forces us to rethink the way we view and design with preexisting structures, avoiding the easy option of starting from zero.

Respectfully using what already exists, but with the freedom to develop a plan that meets new needs

Projects that entail working with preexisting elements can become opportunities to provide unique and innovative solutions.

Making the most of existing elements in the proposed new composition

The number of elements preserved will be directly proportional to the amortized carbon footprint. The more elements preserved, the better. Maintaining them will give the building an identifiable character which, by means of these elements, will explain part of its previous history.

Doctors Without Borders Headquarters in Barcelona (Spain, 2017-2019): a new life for an obsolete building.

3.70 tn CO_2eq/m^2 emission savings with refurbishment

1,400 m^2 of outdoor space promote users' health and well-being

The proposal for the new Doctors Without Borders headquarters in Barcelona transforms an existing building in the 22@ district by promoting spaces that are functional, biophilic and people-friendly. The existing building was an indus-

trial site with a ground floor, an additional four stories and a rooftop level that had undergone various transformations while it was operational. The most important was in 2003, when a complete refurbishment made the former industrial premises suitable for office use. The communication centers and services were modified to meet current regulations without considering the resulting conditions in the workspaces. The building occupied nearly the entire lot, creating deeply recessed spaces with little natural light and ventilation; the rooftop was used strictly for the building's facilities.
The project incorporates different strategies to improve the building's current condition. They not only respond to the program and operational requirements but also study the impact of the different actions and choice of materials, prioritizing a sustainable, environmentally-friendly design committed to a paradigm change with the climate emergency in mind.

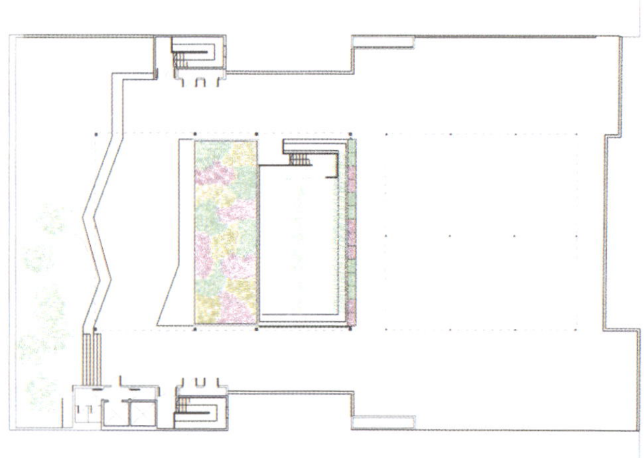

Roof floor plan

Type floor plan

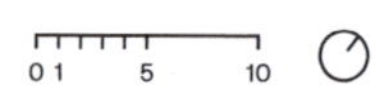

A green terraced courtyard will improve the lighting and ventilation of interior spaces. The courtyard also includes a new set of stairs that reinforce the connection between the floors, thereby promoting healthy habits.

Refurbishing the existing building has resulted in less than half the CO_2 emissions of tearing it down and rebuilding it.

Designing a biophilic building where the rooftop vegetation, façades and interiors are at the forefront, improving people's health, air quality and biodiversity in the city.

Recycling the building's rainwater for use on site, promoting the circular economy. Rainwater is collected from the photovoltaic pergola and stored in tanks on the roof to water the building's vegetation.

Promoting sustainable mobility by situating bike parking at the main entrance of the building; in this way the project complements the city's new sustainable mobility network. Healthy interior routes are also promoted through the inclusion of the new staircase in the central courtyard where it is visible from all points.

Placing strategic elements on the roof such as mobile awnings and cooling surfaces with abundant vegetation that help mitigate the heat island in cities.

Promoting local, environmentally friendly and recyclable materials. Certified wood is promoted as the project's main material.

n/ZEB

Developing passive strategies to achieve minimum energy demand, such as reinforcing insulation on roofs and façades, and replacing all door and window frames to improve energy efficiency.

Connecting the building to the Districlima grid. The rooftop photovoltaic pergola produces electricity, boosting the building's energy self-sufficiency and productivity.

Promoting users' health and well-being is the main strategy of the project, with actions that improve spatial, physical and material comfort such as the incorporation of the central courtyard and the use of natural materials.

n/ZEB

LED lighting with light sensors to optimize the building's power consumption

Colonization of the roof, generating new green outdoor spaces

Movable awnings prevent the roof from overheating in summer, contributing to heat island mitigation

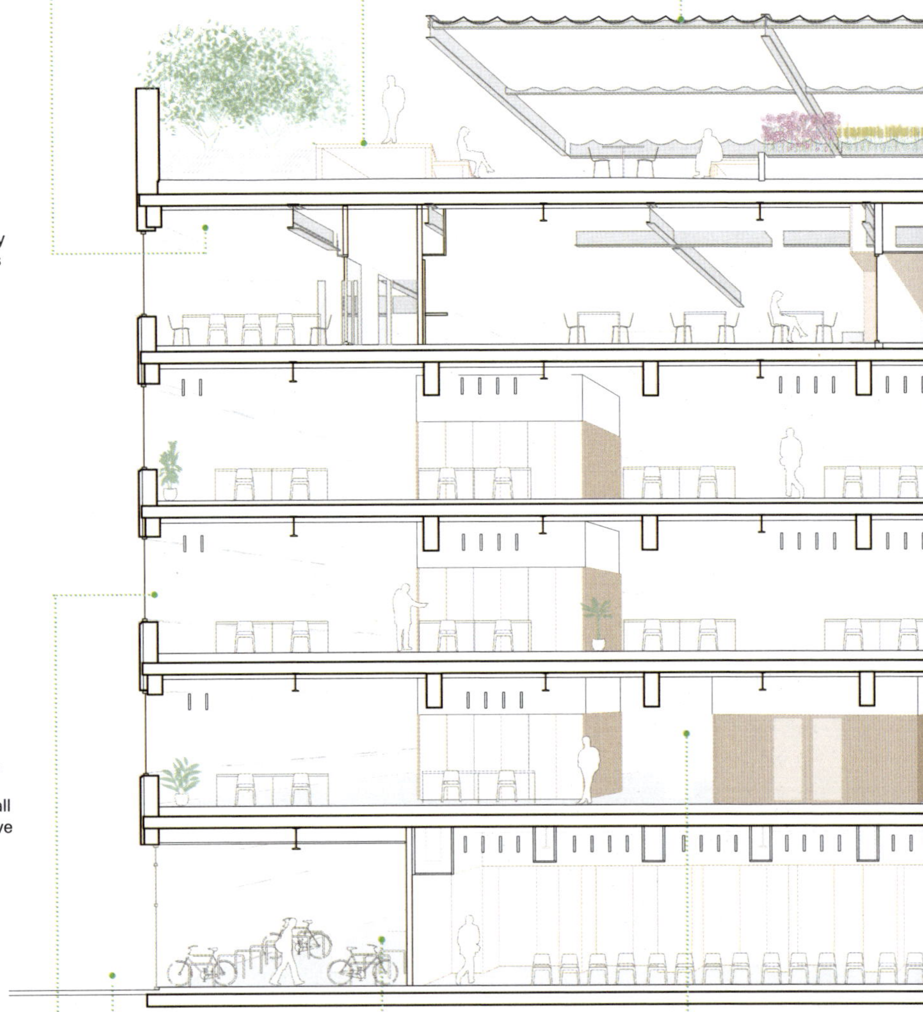

Subway stop 20 meters away

Renovation of an existing building. Emission savings of 3.70 tn CO_2 eq/m²

Bicycle parking next to the building's entrance to promote sustainable mobility

Healthy workspaces with natural light and ventilation

Section. Strategies to combat the climate change

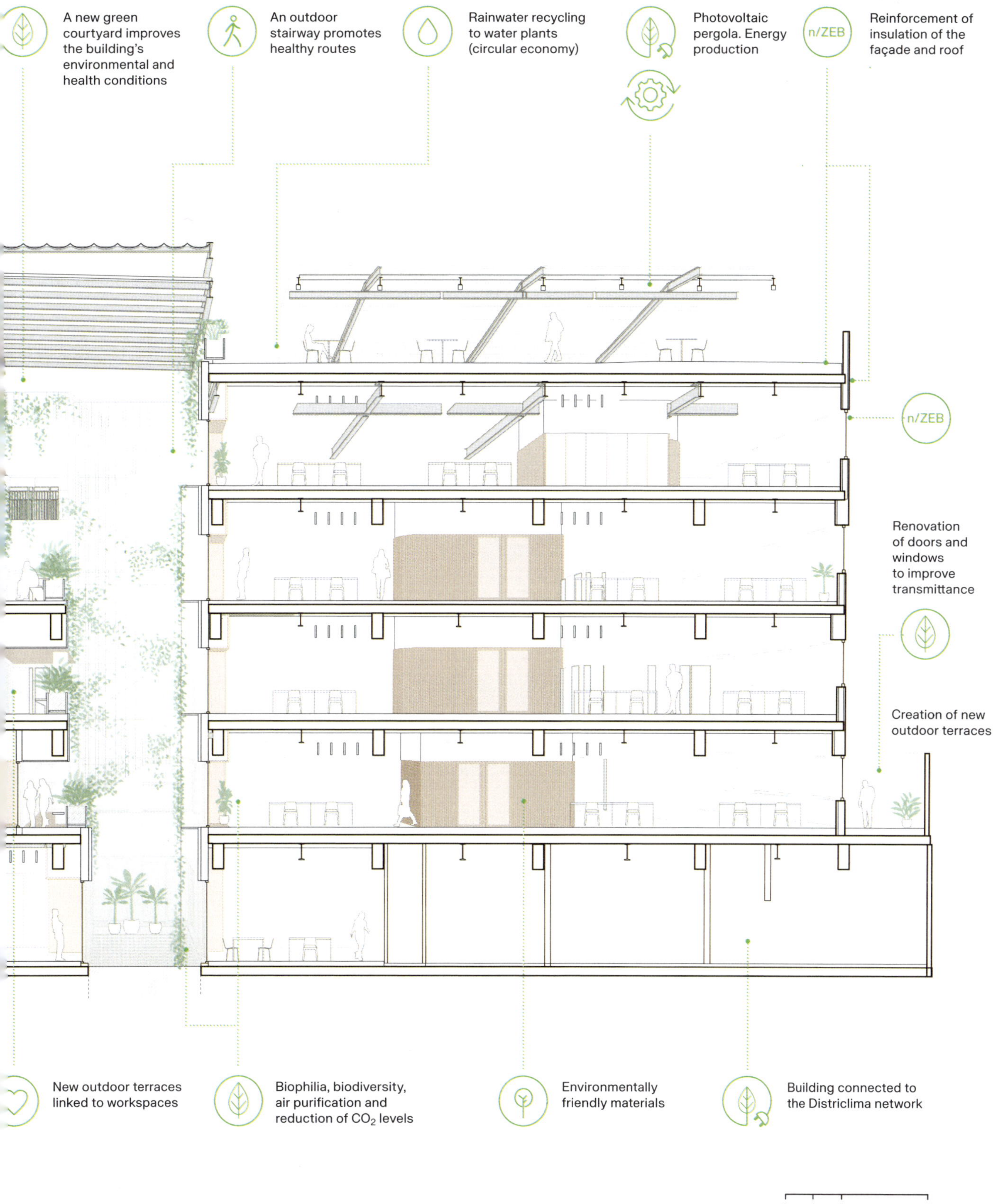
A new green courtyard improves the building's environmental and health conditions
An outdoor stairway promotes healthy routes
Rainwater recycling to water plants (circular economy)
Photovoltaic pergola. Energy production
n/ZEB
Reinforcement of insulation of the façade and roof
n/ZEB
Renovation of doors and windows to improve transmittance
Creation of new outdoor terraces
New outdoor terraces linked to workspaces
Biophilia, biodiversity, air purification and reduction of CO_2 levels
Environmentally friendly materials
Building connected to the Districlima network
0
1
2
5

Building

Metges Sense Fronteres
Ärzte Ohne Grenzen
Bicycle park
Parking bicicletas
Parking bicicletes

The rooftop promotes new outdoor spaces for interaction and work. It transforms a roof intended for the building's facilities into a rooftop meeting place where vegetation is the protagonist. Photovoltaic panels are incorporated into the existing metal structure and rainwater recycling tanks are installed to water plants.

A refurbishment is an opportunity to transform obsolete or inefficient buildings with the aim of giving them a new life, optimizing resources and building with less environmental impact.

On the left page:
(centre) Flexible workspaces. Improved workspace conditions increase spatial sensation and enhance light and ventilation conditions.
(bottom) Ground floor, extension of the street. A visible public pedestrian entrance encourages sustainable mobility over car use.

Client MSF - Médecins Sans Frontières (Doctors Without Borders) / **Authors** Enric Batlle, Joan Roig, Albert Gil / **Team** Miriam Aranda, Annabel Barba, Joan Batlle, Diana Calicó, Elisabeth Torregrosa, Dolors Feu / **Collaborators** STATIC – Structural Engineering, PGI Engineering – Facilities Engineering, Dos54 – Façade Consultancy / **Contractor** Vesta Rehabilitación / **Gross floor area** 8,230 m²

The Search for Nature-Based Solutions

In his book *Cities For A Small Planet*, Richard Rogers proclaimed that sustainability would become the dominant philosophy of our time and that cities, the human habitat, could once again be interwoven with the cycle of nature.

A sustainable city is a city where you can live and breathe, work and walk. Some cities around the world have begun to set themselves the goal of becoming carbon neutral: green mobility, reduction of CO_2 emissions, total recycling of waste, improvement in river and sea water quality, rainwater harvesting, urban energy production, increase in urban biodiversity and, of course, countless public space and building projects modeled on nature-based solutions.

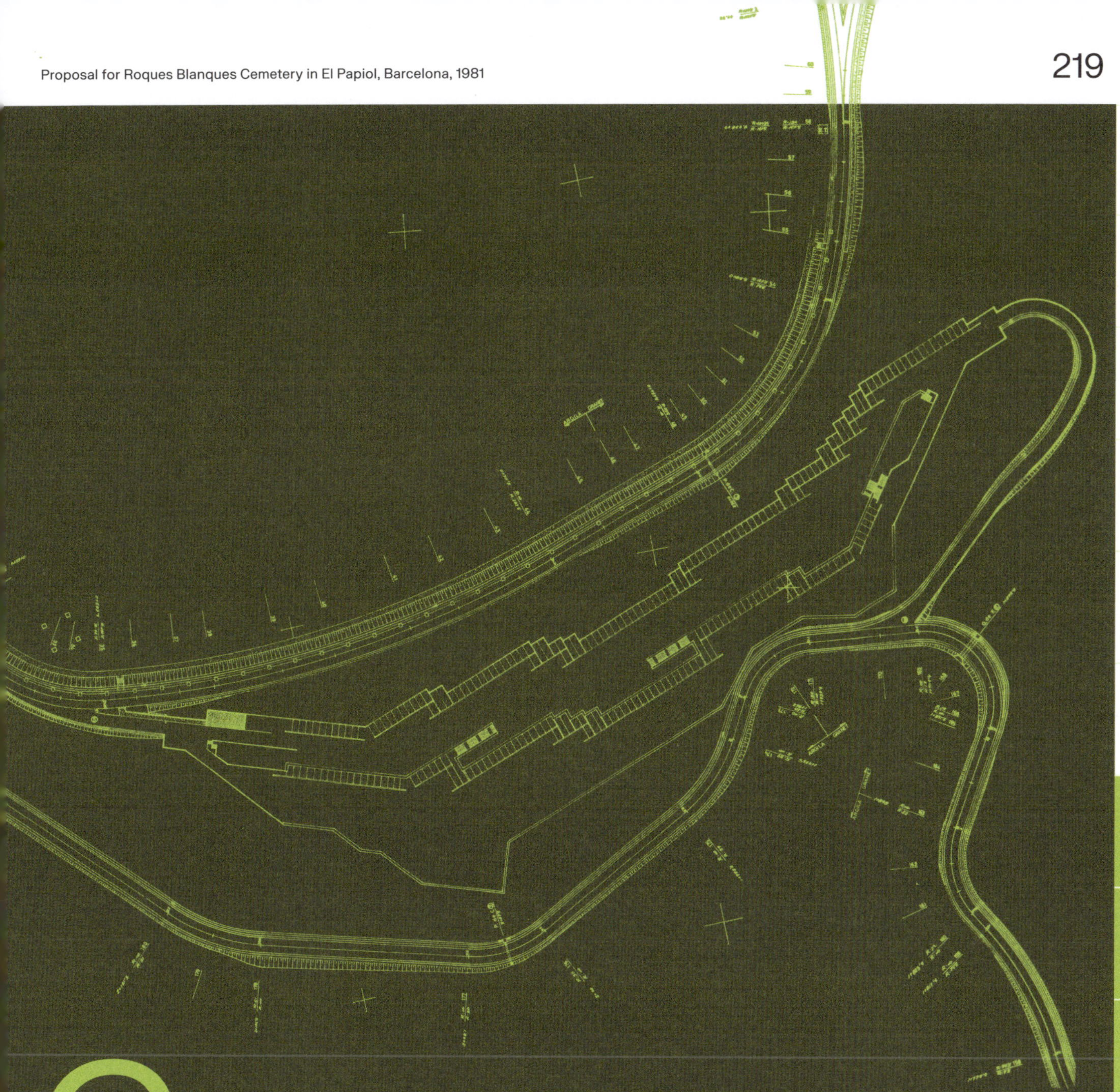

Green Life Cycle

City of Cities

Nature-Based Solutions

Carbon Neutral Balance

City of Cities

In our imagination, the end of a material's life cycle is easy to understand. From a used object, previously considered waste, we obtain the material needed for a new product that we can use. But how do we apply this philosophy to the city? Urban areas today interact constantly with the environmental matrix we want to enhance, and the large infrastructures often located between them. The design of these boundaries, which to many seem residual and unusable, can be strategic, giving a new life to these places by implementing quality uses and creating connections to green infrastructure.

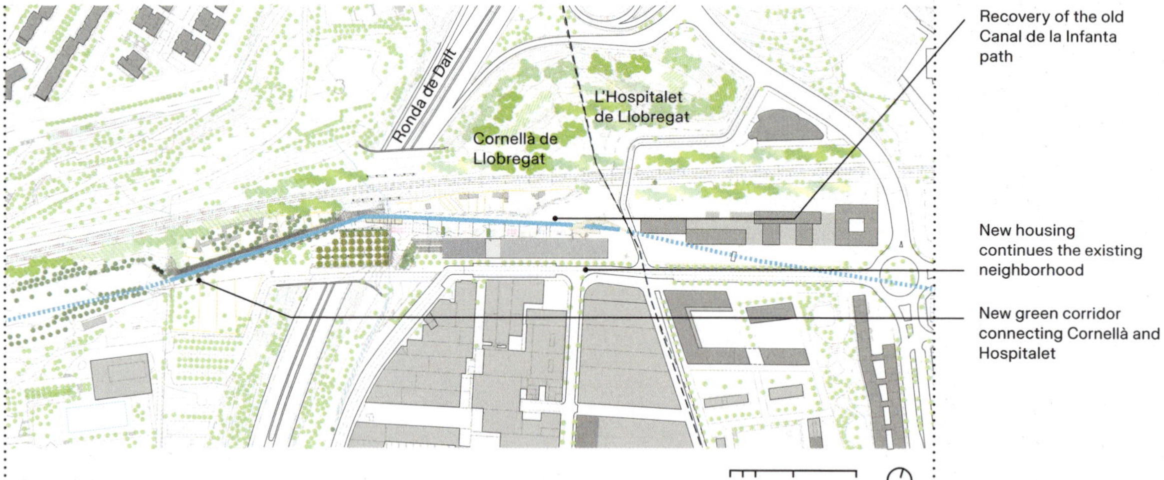

General plan

The Project to Transform the Canal de la Infanta Path reclaims historical agricultural infrastructure that remains on the margins of large transport networks to reunite Hospitalet and Cornellà de Llobregat. It qualifies as a border project in several ways: it is located on an administrative boundary, at the edge of several infrastructures, and on the border between urban fabrics with different uses.

Working at the boundary between the compact city and the environmental matrix

The boundaries between the city and nature, traditionally considered unfinished and pending further expansion, require specific approaches with the aim of permanence that facilitate permeability and exchange between urban areas and nature.

Reclaiming the borders of major road infrastructures

Conventionally designed major transport infrastructures functioned as a barrier and disrupted the urban fabrics they crossed. We need to transform these infrastructures into central places that take advantage of the possibilities offered by their connectivity and unite the urban fabrics to either side.

Redrawing the boundaries between urban fabrics with disparate uses

Improving the transitions between industrial and residential fabrics can create major opportunities for urban transformation and redevelopment, turning these spaces into areas that can become new focal centers in cities.

Gallina Blanca Block Development in Sant Joan Despí, Barcelona (Spain, 2016-ongoing): a new urban centrality around a new modal interchange – train, subway, tram, park and ride.

5 minutes less per trip for residents of the east side of Sant Joan Despí who walk to the town center or take the train

33% of the area for open public spaces

Despite its prime location next to the train station between Barcelona and Molins de Rei, the obstacle in the form of the railway, on the one hand, and the urban development model from the second half of the 20th century, on the other, relegated the block where the former Gallina Blanca factory was located to a marginal position, trapped between the back of the infrastructure and Fontsanta industrial park. The transformation project for this 5-hectare enclosed facility with the modification of the Metropolitan General Plan identifies an opportunity to reverse this condition and transform the space into a new focal center for the city.

The proposal and surrounding urban fabrics, 2022

To this end, the project creates a new street-level passage beneath the railroad tracks and permeabilizes the premises to give continuity to neighboring urban fabrics, extending Passeig Canalies to the other side of the tracks, thereby establishing a natural connection between the old city center and neighborhoods to the east. This new axis, which constitutes a large, equipped space – the new station square – unites around it a new hub of mixed-use activities (free-market and subsidized housing, community facilities, commerce, offices and other economic activities, with the new Agroliment Group headquarters as the driving force) and a new intermodal station, prioritizing pedestrian and bicycle access from both sides of the railway line and eliminating the previous asymmetry between an accessible side and a residual rear part.

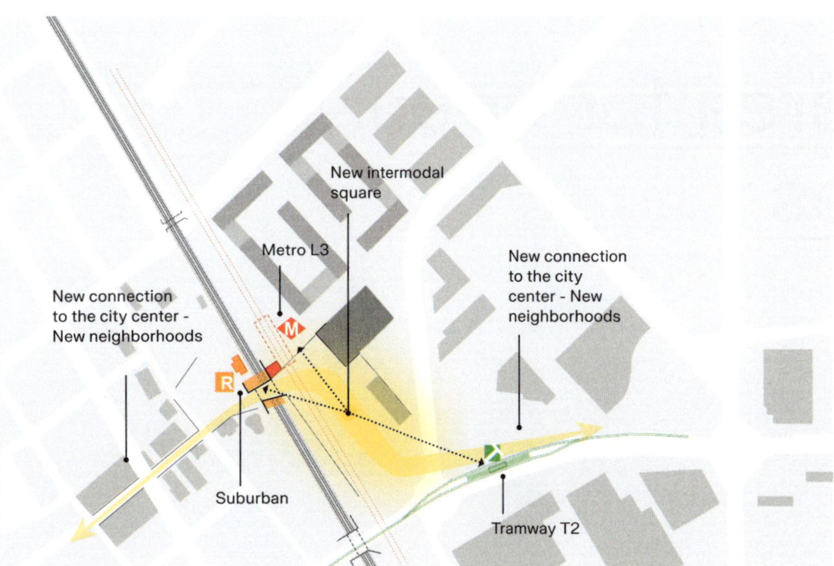

Proposed scheme of the connection between means of transport

The site retains the positive characteristics it possessed when the factory was built, a strategic place close to the railroad tracks and the urban center, even though, over time, the block has become a residual space. The growth of the town and the metropolis has reinforced these original conditions. Consequently, the transformation of the Gallina Blanca block area by means of the modification of the Metropolitan General Plan will convert it into an opportunity, designing a new center that can weave together different urban fabrics.

The new station plaza

Connection of Avinguda de Barcelona with the new intermodal station in the new plaza

Promenade under the railroad tracks

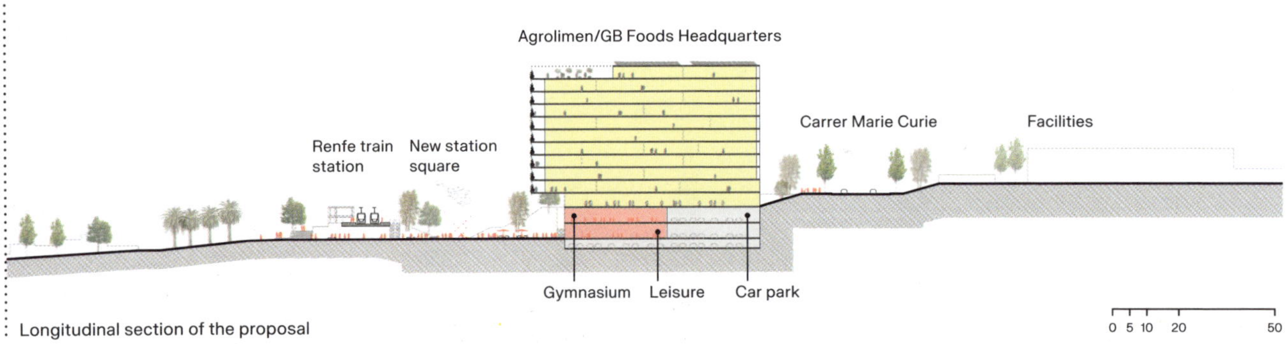

Longitudinal section of the proposal

This mixed-use center of activity is supported by a new intermodal station that combines the existing suburban train, tram and bus lines, and the future extension of Line 3 of the subway, and prioritizes direct pedestrian and cycle access from both sides of the railway line.

Renovation of the suburban train station, placing it adjacent to the underpass and symmetrically with the tracks, eliminates its previous front-back layout, improves accessibility and reinforces the connectivity of the new underpass.

Extension of Passeig Canalies with a crossing under the railroad tracks

Creation of a new centrality

The new axis is laid out in the form of a large equipped public space. The new plaza in front of the station is central and spacious, receiving pedestrian flows from neighborhoods to the east. The square is structured around a mixed-use arrangement that combines free-market and social housing, community facilities, commerce, offices and other activities.

The new focus of activity is a seam joining urban fabrics with different uses. The obstacle presented by the railroad tracks between the old town and urban expansion to the east disappears with the creation of a crossing beneath the tracks from the station square and with the opening of the Gallina Blanca area, transformed into a permeable space that enables the continuity of neighboring urban fabrics.

Client AIRINA / **Authors** Enric Batlle, Joan Roig, Abel Porcar. Lluís Cases (Garrigues) / **Team** Ronda Quetglas, Mercedes Blay, Helena Valls, Ignasi Rincón, Iván Sánchez / **Collaborators** Garrigues – Lawyers, Duatis Arquitectes – Economic and financial study, Cinesi – Transport consultancy / **Project area** 60,717 m², 87,634 m² (Gross floor area)

Nature-Based Solutions

In contrast with traditional environmentalism based on the three Rs – Reduce, Reuse and Recycle – to slow down the depletion of natural resources, in recent years the focus has been on solutions that generate a closed waste-free life cycle with a positive balance. This is the cradle-to-cradle concept, according to which, in nature, everything is useful and forms part of an eternal cycle related to solar energy called upcycling.

Parc de Vallmora in El Masnou, developed from reclaimed farming terraces, was designed using nature-based solutions that maximize the biodiversity of the whole. It also is committed to the playable city, integrating a series of play facilities built with nature and for nature.

Building with nature

We can design landscapes based on the dynamics of nature, optimizing resources and increasing system resilience. Bioengineering techniques that use natural and local materials and techniques recreate the *modus operandi* of nature and evolve with it.

Building for nature

We must give up the pretense of taming the environment. Instead, we can aspire to merge with it without relying on expedients that distort its inherent harmony. We need to develop new elements that not only meet our needs but also have a second common objective: to connect human activity to nature.

Rethinking the way we do things

Each new project can be designed using more natural techniques and materials, to become a space that provides new links with the natural cycles of the environment. Every place, every solution adopted, every choice made can be important and contribute to creating a new green life cycle.

Forest Path in Roques Blanques Cemetery in El Papiol, Barcelona (Spain, 2017-2020): an innovative landscape intervention for a new model of nature-based ephemeral burial sites

1,280 biodegradable burial sites for ashes integrated into nature

2,640 native plants that form a butterfly garden

300 m of terraced land, a new cemetery model integrated into nature

Roques Blanques Cemetery: 37 years on

On July 4, 1981, we won our first competition for the construction of the Roques Blanques Metropolitan Cemetery in El Papiol, an achievement that we consider to be the founding act of our firm. The project involved building a columbarium niche cemetery inside Collserola Natural Park. Our proposal was to create a system of terraces that integrate the niches into the landscape by means of a series of resources: architectural (retaining walls to house the niches), agricultural (cultivated terraces) and gardening (pergolas and hanging gardens). Viewing the Mediterranean cemetery as a garden integrated into the landscape was considered a unique contribution in a context that perhaps placed more value on monumental, architectural and symbolic capacity than new cemetery projects could provide.

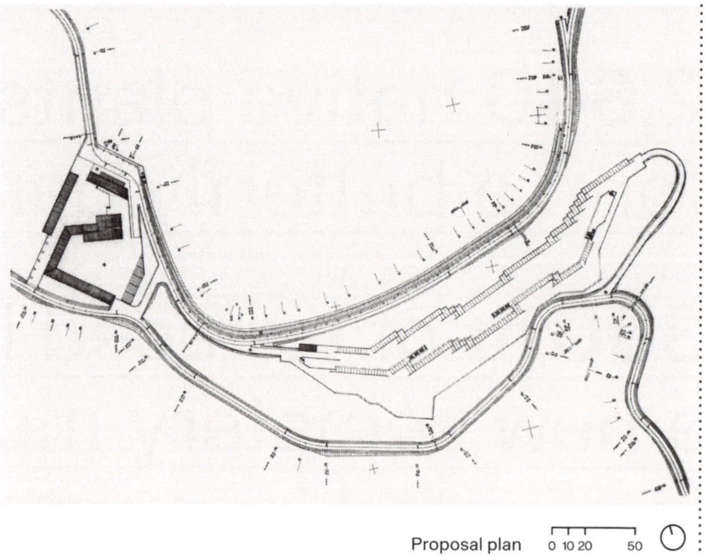

Proposal plan

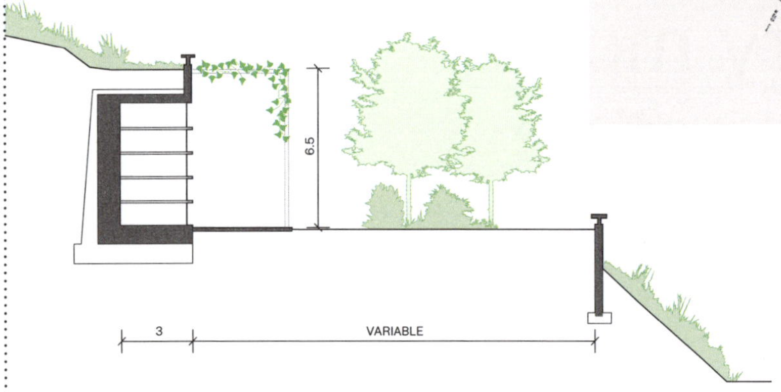

Type section of the Area 1 niches, 1985

In July 2018, we won the Catalunya Construcció Prize for Innovation for the project Forest Path, within the cemetery itself, an example of how we apply the concept of changing everything so that everything stays the same. The project consists of a cemetery for incinerated ashes within a landscape like the one that was there before. The proposal introduces a terraced path that integrates the ashes and promotes the biodiversity of the surrounding area. Once again, we view the Mediterranean cemetery as a garden integrated into the landscape but with two fundamental changes from the original proposal: on the one hand, a type of grave that allows us to work at another scale and with other materials; and on the other, the constructive solution of the terraced land, which allows us gradually to dissolve the architecture into the landscape.

Proposal plan

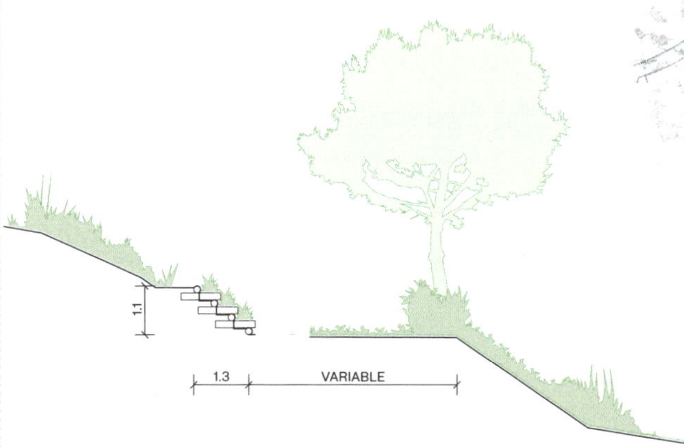

Type section of the Area 6 Forest Path, 2017

Building with nature
The place where the ashes are deposited forms a soil retaining wall. It is a modified space that combines dead and living material: the living formwork as a 100% biodegradable alternative that adapts to the natural dynamics of the forest. We learn from beavers, who build their shelters – feats of environmental engineering – from tree trunks, branches and dirt. The organic urns for the entombed ashes are made of chestnut wood and located in spaces within the formwork, resulting in a construction with its own integrated materials, as part of the natural biodynamics of the forest.

Building for nature
The place where mourners bid farewell to their loved ones is transformed into a Butterfly Garden. This biodiversity-enhancing area revitalizes useful spaces in Collserola Natural Park. It is a new biotype for insects and small animals where plant roots anchor the soil and retain moisture, and plants and flowers provide the food.

Forest pantheons. Plant apses designed to house underground graves surrounded by nature.

Biodegradable burial space

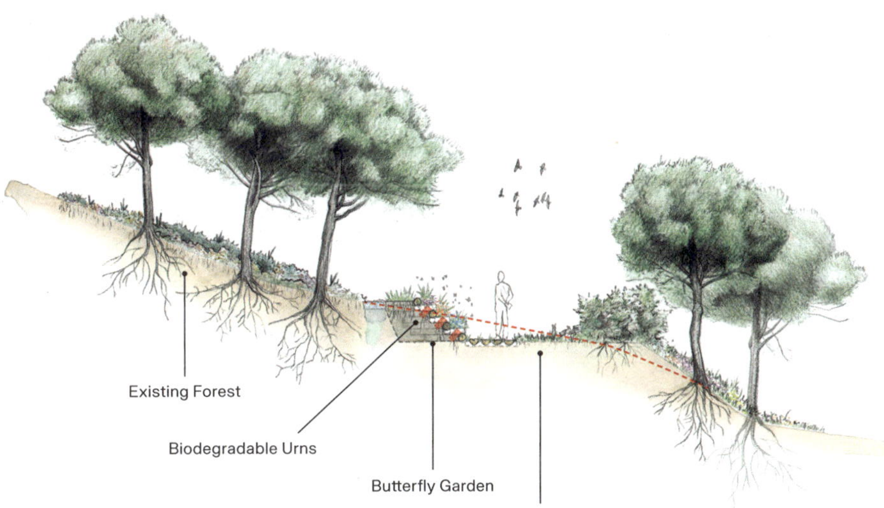

Type section of the Forest Path, Area 6 - 2017

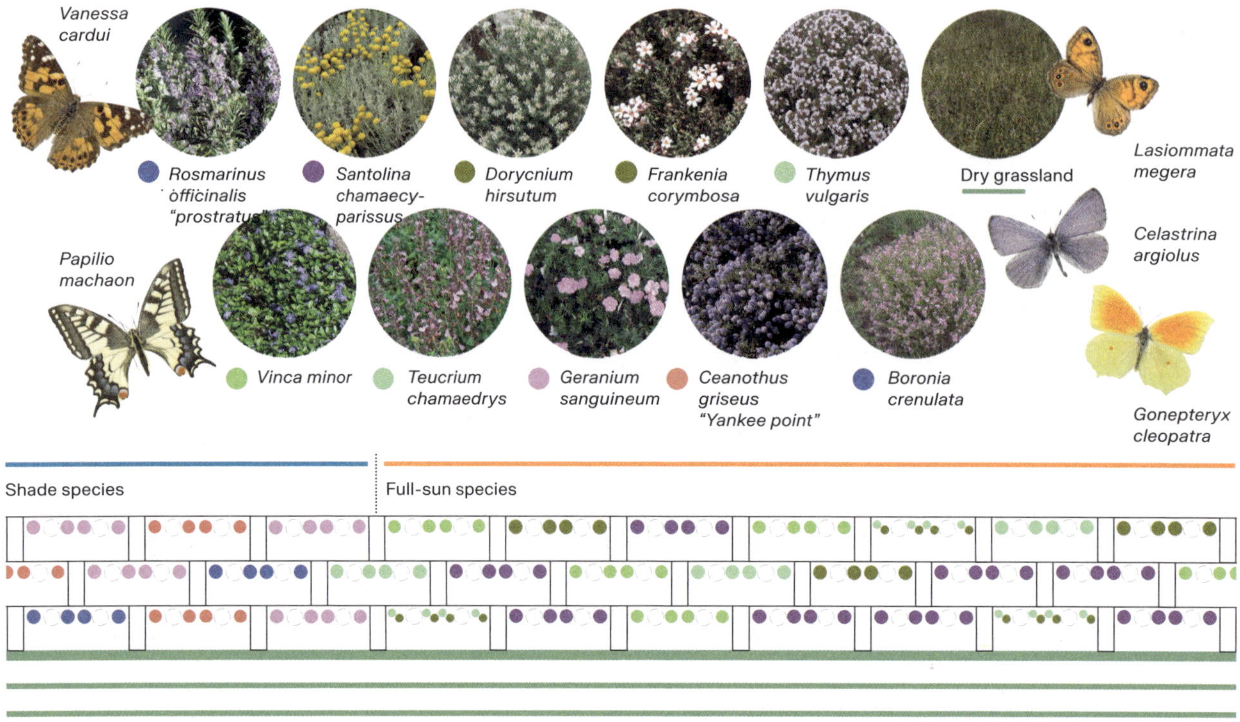

Shrub and herbaceous species with brightly colored and contrasting blooms to attract butterflies.

The place chosen for the Butterfly Garden is a Krainer wall, a space shielded from the wind, one of the main enemies of this insect. Furthermore, it is intensely sunny, ensuring a generous, long-lasting blooming time. The vegetation is both native and adapted, combined in a way that does not create competition and enhances its ornamental qualities.

To encourage the presence of butterflies, we chose shrub and herbaceous species with brightly colored blooms, able to attract these insects and with good nectar production as the primary food for adult butterflies and a chewable leaf for those in the larval phase (caterpillars).

Memorial, Montserrat lookout. A place for saying farewell to the deceased.

Client GIC – Áltima / **Authors** Enric Batlle, Joan Roig, Iván Sánchez / **Team** Marta Sanz, Livia Valentini, Carmen Guerrero, Elisabeth Torregrosa, Laura Luque, Dolors Feu / **Collaborators** Naturalea – Consultant in bioengineering and landscape rehabilitation techniques, Cave Terram – Environmental consultant, WorldNature – Consultant in nature conservation and environmental consciousness / **Contractors** Naturalea, Excavacions Petit, Santin Jardineria i Paisatge / **Project area** 8,909 m²

Carbon Neutral Balance

Carbon neutral balance aims to achieve zero carbon dioxide emissions. This is achieved by balancing the amount of carbon dioxide released into the atmosphere with the equivalent amount removed. The strategy for putting it into practice consists of abolishing the concept of waste and promoting local recyclable materials with minimum CO_2 consumption in their production.

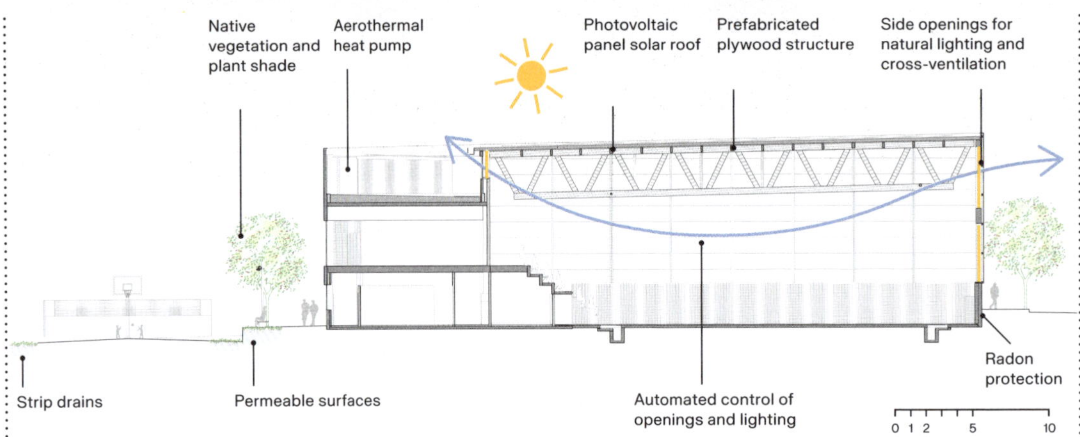

In the plans for **the Sports Pavilion in Reus**, the application of this life cycle analysis methodology led us to modify project solutions (substitute the steel structure with wood, remove false ceilings, substitute drywall with thermo-clay...). All these solutions allowed us to move from CO_2 emissions of 378 tn to 49 tn, representing an 87% reduction.

Cradle to Cradle

The concept appeared with the book *Cradle to Cradle: Remaking the Way We Make Things*, written by the ecological chemist Michael Braungart and the American architect-landscape architect William McDonough. The book represents a new way of interpreting environmentalism, where the materials complete a closed circle "from cradle to cradle" from which the concept of waste disappears. Everything is used and returns to nature or has another use.

Analyzing the life cycle to estimate and assess environmental impact attributable to the building

Only by quantifying its impact can we assess our CO_2 footprint. To do so, we must incorporate life-cycle measuring systems into projects, the same way we measure amounts of materials or construction costs.

Rethinking business-as-usual building solutions and applying materials and systems that reduce or eliminate our carbon footprint

Only by changing can we truly reduce. The new systems will involve new ways of thinking and designing to ensure these solutions are implemented in the most favorable and logical way. These systems will promote the use of materials such as wood with closed life cycles.

La seguridad crea la libertad.

Vega Instrumentos Head Office in Sant Cugat del Vallès, Barcelona (Spain, 2019-2022): the first private office building in Spain with an all-wood structure.

650 tn of CO_2 stored in the building structure and not released into the atmosphere

52 trees that capture CO_2 as they grow and 3,300 m^2 of green areas

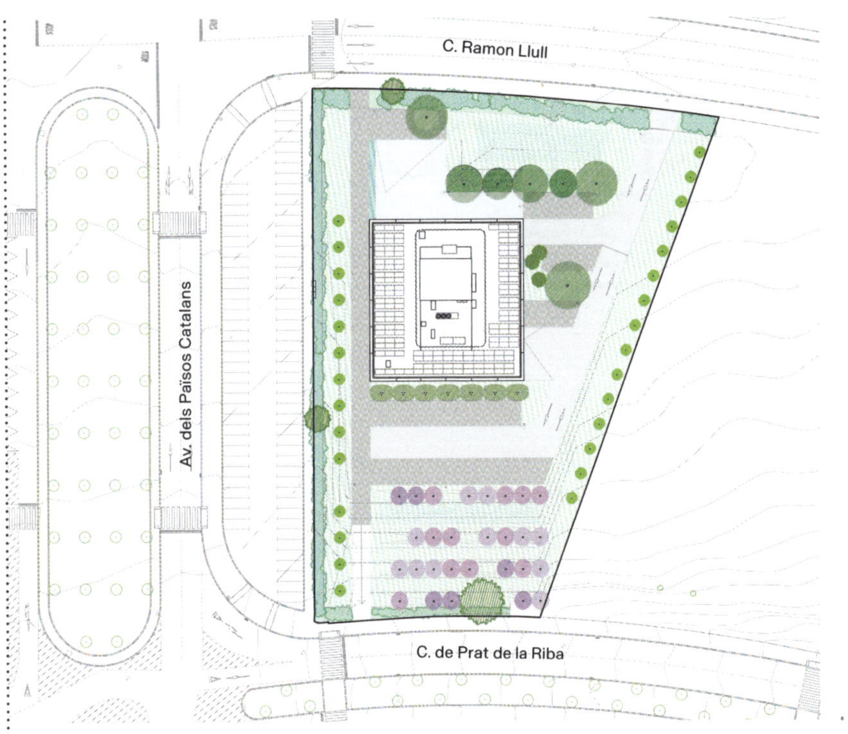

Location

The new Vega headquarters consists of two 30 x 30-meter floors with a 5 × 5 modulation. This construction approach ensures easy extension of the building and maximum efficiency. The public ground floor, which will be used to receive visitors and as a showroom and area for repairs, is designed as a more compartmentalized space. Meanwhile, the upper floor, for internal use and where the work-spaces and offices are located, has a more open-plan layout.

A modulated façade with a high degree of insulation and air-sealing that minimizes energy consumption

The enclosure has a ventilated façade system with glazing equivalent to 30% of the total façade, equipped with automated horizontal louvered shutter systems that achieve an optimal balance between user comfort and the building's energy efficiency.

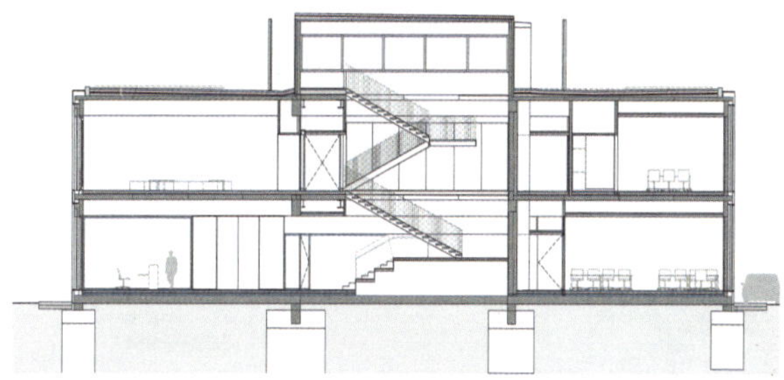

Section

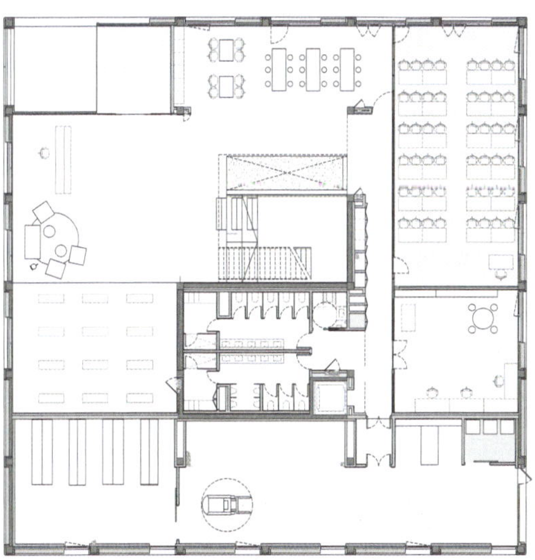

Ground floor plan

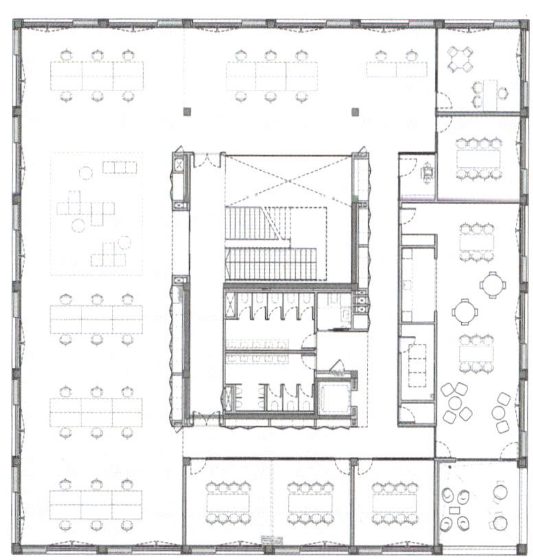

First floor plan

0 1 5 10

Promoting sustainable mobility in the building
The building has an open staircase that encourages use: easy to locate and intuitive, with natural light, ample space and vegetation. The overhead opening allows natural light to enter the central part of the building.

Healthier, more domestic workspaces
The combination of exposed structural wood, along with the more human-scale window system, white acoustic cladding and textile floor coverings, result in interiors that provide warmth and well-being.

Wood: the protagonist
The structural system is based on the use of GL24H laminated timber portal elements on which the CLT floor slabs are placed.

It is a logical system with sensible spans of 5 × 10 meters. Insignis pine wood (*Pinus radiata*) from the Basque Country is used, processed there and transported to Sant Cugat, opting for a local product with the minimum possible transport. As a result of this choice, 650 tonnes of CO_2 are stored in the building's structure.

Client VEGA Instrumentos / **Authors** Enric Batlle, Joan Roig, Albert Gil, Helena Salvadó, Marcos Andrés / **Team** Clàudia Amías, Laura Luque, Gemma Ramon / **Collaborators** G+R Structures – Structural engineering, Estudis ESL – Facilities engineering, Egoin – Engineering, manufacturing and construction in wood, GPO – Execution Management / **Contractor** VOPI4 / **Gross floor area** 1,800 m²

n/ZEB

Innovating to Reduce

We have worn out the word sustainable before we could rigorously put it into practice over a long period of time in a meaningful space. According to Joaquín Araújo, sustainability is an attempt to insert our activities in an open process that recognizes origins and does not destroy the unknown destination. To speak of logical sustainability is to apply simple, rational solutions to each problem, testing in each project, based on the specific innovation needed, concrete solutions that can help to achieve the objective of reducing emissions.

In his book *Que lo hermoso sea poderoso* (The beautiful must be powerful), Ramon Folch contrasts the environmentalist slogan "small is beautiful" with the developmentalist slogan "big is powerful". Perhaps now is the time to advocate a third way, removed from neo-ruralist environmentalism and industrialism: if what is small is beautiful and what is big is powerful, it is reasonable to propose that what is beautiful be powerful, that what is "eco-logical" be useful and universal.

Near-Zero Emissions

A Naturally Regulated Environment

Logical Sustainability

Near Zero-Energy Buildings

A Naturally Regulated Environment

City design that seeks zero emissions must begin with urban planning, with strategies that take advantage of the natural characteristics of the place, the design of the urban form and the regulations that promote the introduction of energy efficiency criteria throughout the city. The measures that urbanism can introduce, because of their essentially passive nature, are undoubtedly ones that can have the most impact on improving the energy efficiency of the places where we live.

The urban extension proposal in **the Passeig Nord Area, in Reus**, was based on the agricultural plan of arranging the built-up areas and open spaces in successive strips. The green infrastructure forms large corridors connected to peripheral agricultural areas and alternates with new urban areas that grow in the form of fingers from the urban center, which enjoys the proximity of the greenery and the natural environmental regulation it provides.

Taking advantage of the geography of the site

We need to understand the geographic characteristics of the site, considering topography, orientation, and natural and climate structures in order to design proposals so that future development benefits from them and does not depend on energy consumption to generate comfortable spaces.

Designing urban patterns to minimize future energy demand

We can contribute to reducing the energy demand of buildings and open spaces during their useful life by designing green areas to serve the maximum number of buildings, considering the orientation and separations between them to ensure spaces with good sunlight and ventilation, or establishing optimum maximum building measurements for their future use.

Introducing regulations that promote passive systems

By means of urban planning regulations, we have to encourage the introduction of passive systems, both in the form and construction of buildings and in the design and configuration of open spaces. We need to promote intermediate spaces, solar protection elements, ventilation courtyards, abundant vegetation, sustainable and local materials, and dry construction systems with less associated waste.

Llevant Mar Sector in Gavà, Barcelona (Spain, 2017-2021): a development that respects and enhances natural values, providing continuity to the biophysical matrix in a privileged environment.

+7 ha of pine forest conservation on public and private land

+2 ha of dune system recovery

65% reduction in areas affected by building compared to previous planning

Existing

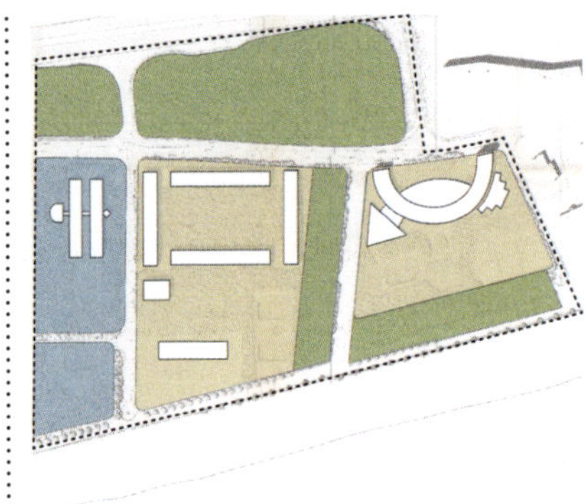

Current 1990 plan
34,500 m² green space
16,024 m² residential housing
10,000 m² facilities

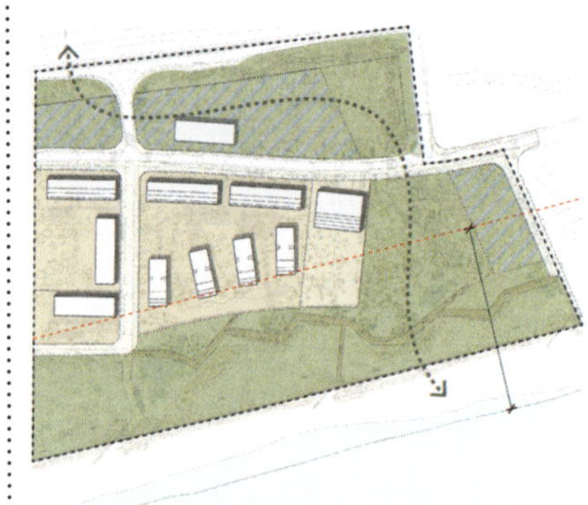

Proposal
70,000 m² green space
24,709 m² residential housing
3,000 m² hotel

Our proposal is based firstly on the premise of eliminating all the existing buildings along the beach and recovering the dune system on this strip, which is either highly deteriorated or, in the built sections, has disappeared; and, secondly, on analyzing the situation in the rest of the sector, which is partly urbanized with natural but neglected growth, taking advantage of the elements inherent to the place.

We propose to maintain and improve at least 80% of the existing pine forest, not only preserving it in green public areas but also in open areas on private land, structuring it to connect the dune system to the interior and the Baix Llobregat Agricultural Park.

Plan of the proposal, showing the recovery of the dunes and the integration of building with the existing pine forest.

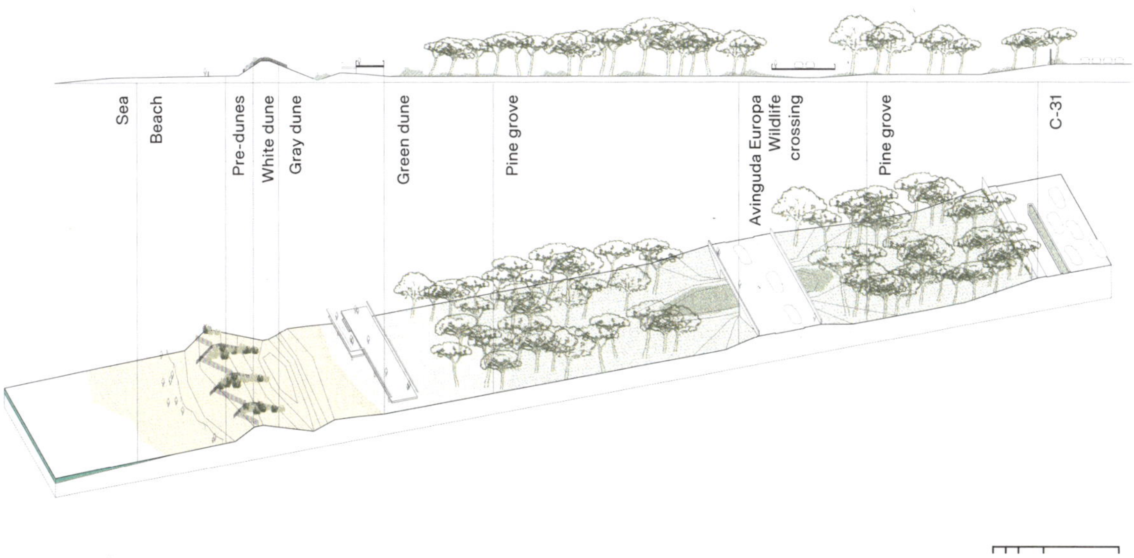

Dune recovery section

To preserve the existing vegetation and intervene in the least aggressive way possible, we are studying how to implement buildings above the natural topography with semi-underground parking lots to reduce necessary earthworks as far as possible, allowing for maintenance and natural runoff. Likewise, the building layout ensures good sunlight and ventilation, with an urban design that supports openness toward the sea, facilitating the passage of the sea breeze and the entry of sunlight from the south, while reducing the energy demand of homes during their useful life. A staggered volume will further promote good lighting in all the homes and on their terraces, as well as less visual impact on the waterfront.

In the urban development of private land, dry construction and sustainable materials such as wooden walkways or floating platforms on the original terrain are prioritized. This preserves natural shade in the pine forest and the delicate ecosystem of dunes while reducing the environmental footprint during construction, maintenance and demolition.

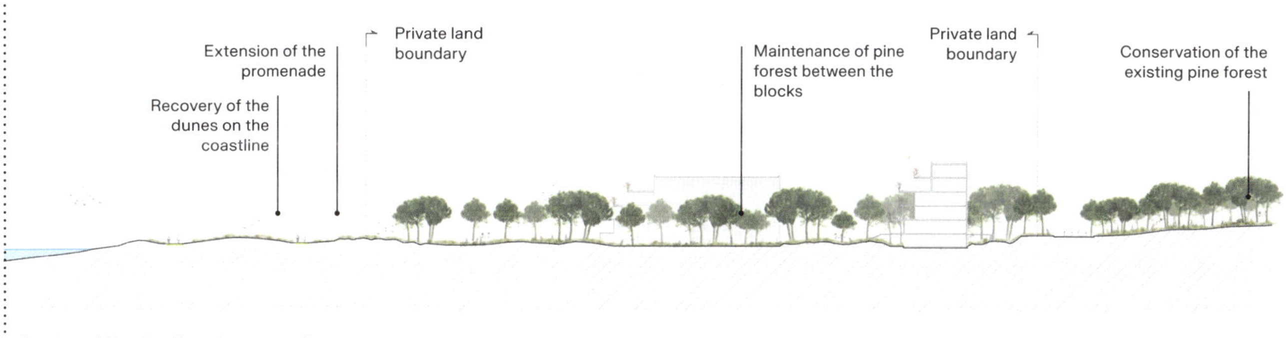

Section of the pine forest conservation proposal between the stilt buildings

Client Urban Input / **Authors** Enric Batlle, Joan Roig / **Team** Abel Porcar, Helena Valls, Joan Batlle, Mercè Lorente / **Collaborators** Ecafir – Environmental Engineering, Garrigues – Lawyers / **Site area** 117,718 m²

City and Territory

Logical Sustainability

We should strive to design landscapes that are self-managing and do not require maintenance that involves an excessive carbon footprint. We need to optimize the use and management of available resources to reduce emissions and improve resilience to the effects of climate change. To this end, projects must be based on a careful reading of the site to enhance its values and be able to dispense with everything that is added and superfluous.

The Access Walkway from the New Parking Lot to the Thau School is 100% wood; this avoids having to alter the topography or remove any trees from the pre-existing slope.

Prioritizing self-management in the landscapes that make up the public space

The goal is to have landscapes that are self-managing or only require simple maintenance actions that minimize the use of machinery and mechanical transport. For this, we must implement solutions with an abundance of vegetation and minimize the hard spaces and excessive equipment commonly found in our public spaces.

Optimizing resources and taking advantage of pre-existing assets

Projects need to begin with the reuse of existing elements and materials at the site: an agricultural wall, a tree, topsoil, pavement, shrubs, a path, and especially water and its various routes.

Eliminating what is added and superfluous

Modulating intensity in the actions we take serves to optimize the resources used, reducing emissions and improving the resilience of public spaces.

Igualada Central Park, Barcelona (Spain, 2017): a large, landscaped amphitheater that establishes an important approach route to the city as a green corridor and social connector between the city and its natural surroundings.

100% dry construction

600 m of wooden stands that create new meeting and leisure spaces for the inhabitants of Igualada

100% permeable soil

50% shade for the new amphitheater with the planting of 67 trees

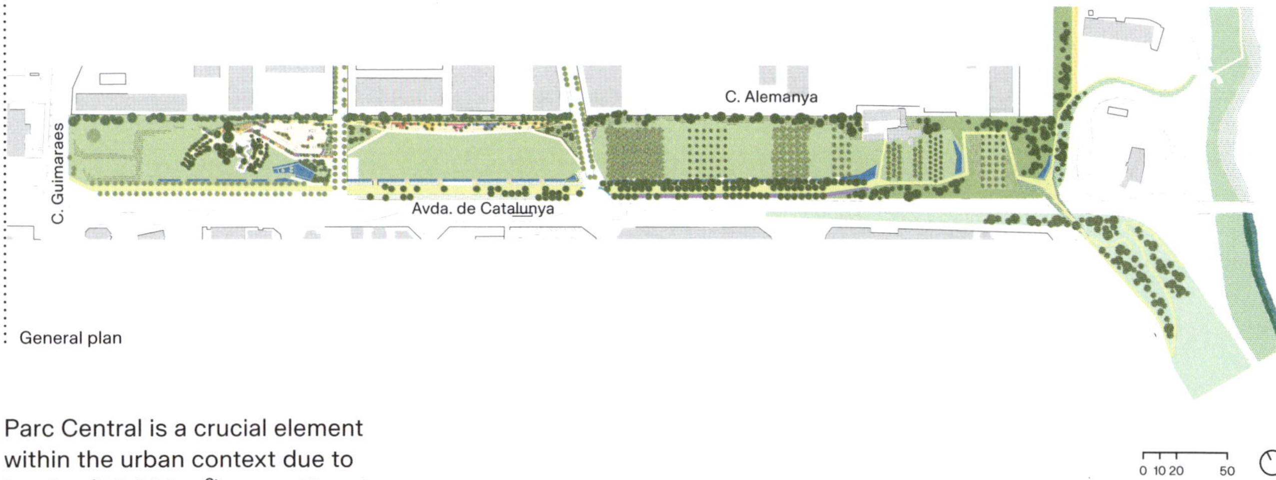

General plan

Parc Central is a crucial element within the urban context due to its size (80,000 m^2), as a pillar of Igualada's open-space strategy: a large green area that connects the interior of the city with its natural surroundings.

The main objectives of the proposal are to strengthen the role of the green corridor, contributing to the city's biophilization; commit to its potential as a social connector between the city and its natural surroundings, bringing the richness of the natural spaces closer to its inhabitants; and design environmentally friendly solutions that embrace the multipurpose function of open space.

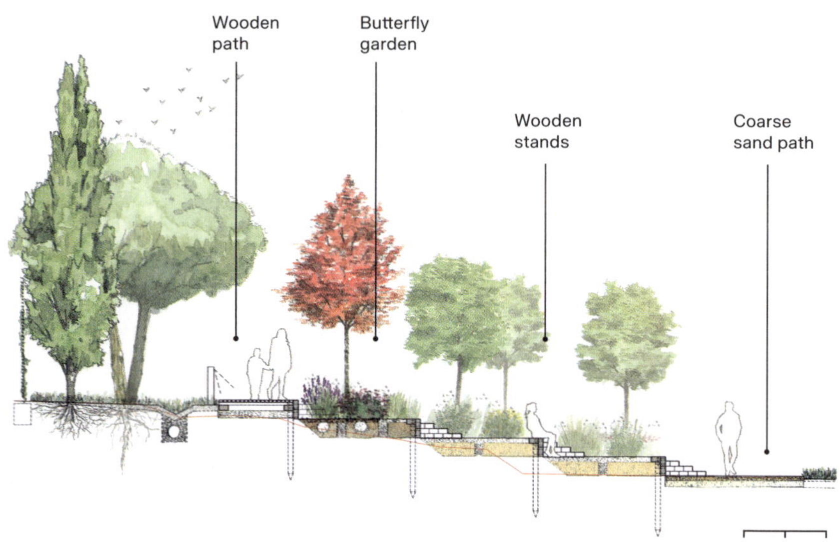

A resilient garden

The terrace system has a dual function as a place to sit down and a large garden. The garden improves bioclimate conditions, providing shade and increasing biodiversity in the city.

Topography as the first tool

The project maintains the large central grassy area as a multipurpose green square for ephemeral uses and promotes healthy and recreational activities in contact with nature while simultaneously focusing its efforts on improving the unevenness between the park and the industrial area that bounds it on one side. The existing slope has a difference in height of approximately 2 meters, used to create a linear amphitheater with a capacity of up to 1,500 people, where the ample grassy space becomes the perfect setting for large-format activities.

Celebration of the European Balloon Festival held every year in the park esplanade

Environmentally friendly construction
The large amphitheater is designed as an agricultural terraced landscape made up of permeable surfaces and low retaining walls to sit on and enjoy the spectacle of nature and different events such as the European Balloon Festival, held every year in the park esplanade.

Client Igualada City Council / **Authors** Enric Batlle, Joan Roig, Iván Sánchez / **Team** Livia Valentini, Dolors Feu, Marta Sanz, Marc Torrella / **Collaborators** SBS Simón i Blanco – Civil Engineering / **Contractor** Acsa Sorigué / **Project area** 6,303 m²

Near Zero-Energy Buildings

By working with passive strategies, we can dramatically reduce the energy demand of buildings. The proper design of buildings, adjusting the treatment of façades according to orientation, with correct solar protection, adequate thermal insulation and incorporating natural ventilation, allows us to meet the requirements of the various regulations (CTE, EPBD, Passivhaus) and become more in line with nZEB criteria.

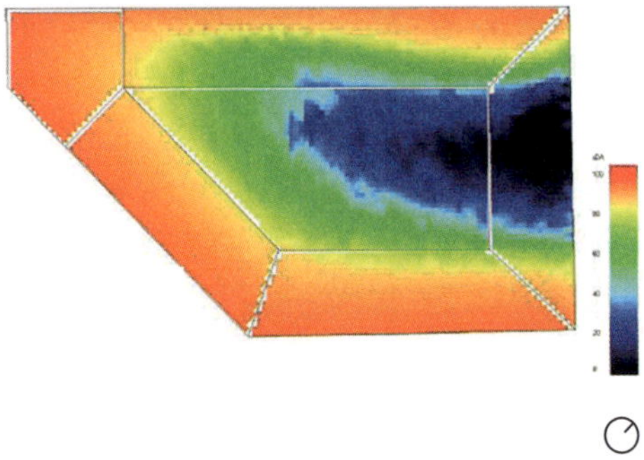

The **Entegra office building in the 22@ district in Barcelona** was the subject of an energy efficiency measures study that reduced net primary energy consumption to 23.9 kWh/m² per year. This value is lower than Passivhaus regulatory standards and the recommendations of the EPBD directive, granting the building an nZEB classification. In turn, it is more than four times lower than the regulatory value required by the Spanish Technical Building Code.

Developing passive strategies to reduce energy demand

It is common to focus on reducing building consumption, improving the efficiency of installations or increasing the contribution of renewable energies. However, passive strategies have a direct impact on demand reduction, and this is the first thing we need to address when designing a building.

Increasing natural ventilation

We must apply passive strategies on façades to reduce buildings' emissions such as: managing them according to their orientation (Window to Wall Ratio WWR, % of openings in the façade), incorporating natural ventilation to generate night and free cooling, controlling air infiltration, increasing their insulation as well as that of roofs, and installing solar protection.

Exceeding the requirements of the applicable regulations to approach nZEB

We have to go beyond the applicable regulations and try to get closer to nZEB classification. For example, the Technical Building Code (CTE) limits the net primary energy consumption of an office building in Barcelona to 100 kWh/m²/year while Passivhaus and EPBD standards respectively recommend a net primary energy consumption of 15 and 30 kWh/m²/year. Even so, technically we have the capacity to lower these figures and to approach nZEB classification levels (buildings that consume practically the same amount of energy they produce from renewable sources on site or nearby).

Tànger 36 Office Building in Barcelona (Spain, 2018-2021): a building integrated into Poblenou's industrial past with a ceramic-toned façade and a large entrance atrium that serves as a climate refuge for the city.

153 m² entrance courtyard transformed into a climatic refuge for Barcelona

1,150 m² of outdoor areas that promote people's wellbeing through the inclusion of biodiversity and vegetation

237 m² of outdoor space for every 1,000 m² of indoor office space

LEED GOLD Certification

Tànger 36 is a state-of-the-art office building on the corner of Carrers Tànger and Àlaba, bound to the rear by Passatge de Ratés, a planning strategy that batlleiroig also implemented.
The building space is organized in an L-shape on the first four floors, plus a technical floor above which comprises two smaller levels. On the fifth floor, where the section change occurs, is a green rooftop with landscaped spaces for possible events and socializing. In addition, a large distinctive atrium in the main entrance creates a climate shelter and improves the building's thermal and acoustic insulation, including shaded and wellness areas for users.

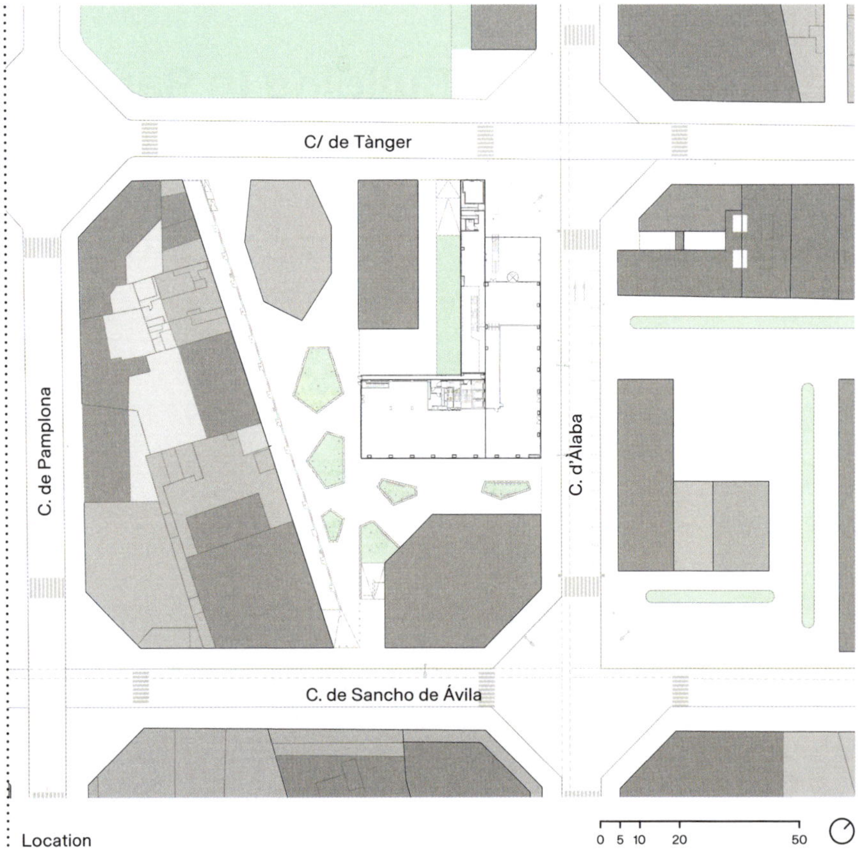

Location

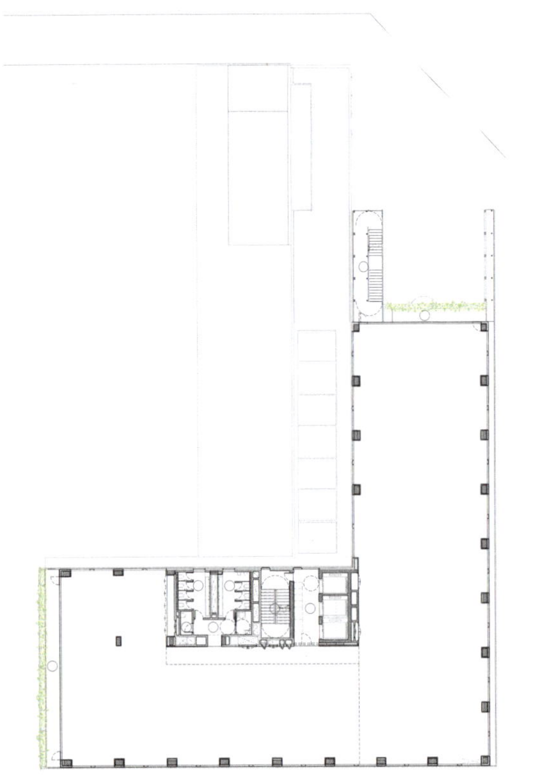

First floor plan

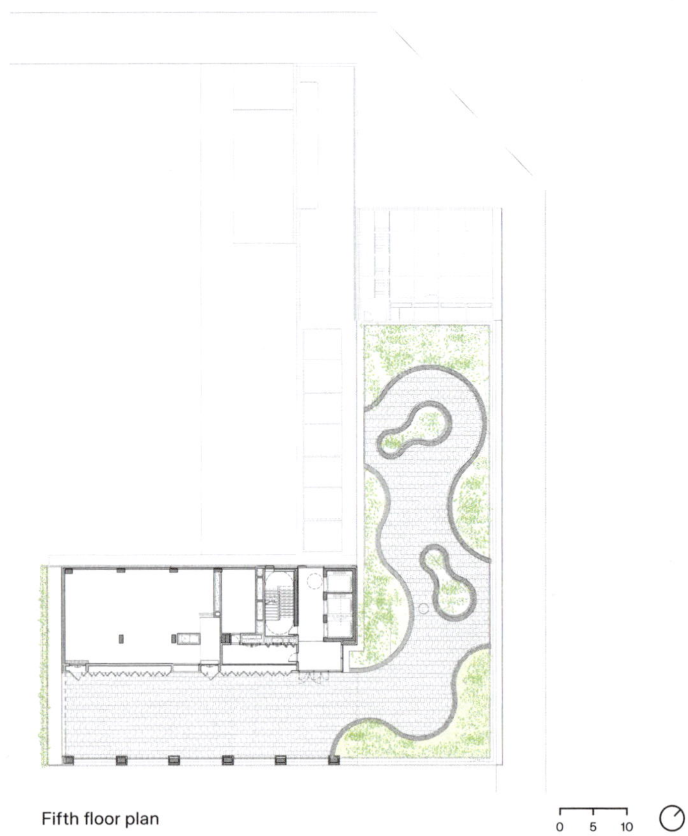

Fifth floor plan

The project design follows optimization and rationality criteria, with the placement of the communications and services shaft in the center of the floor, emergency stairs outside and the entire vertical structure on the façade, resulting in versatile open-plan floors for the offices.

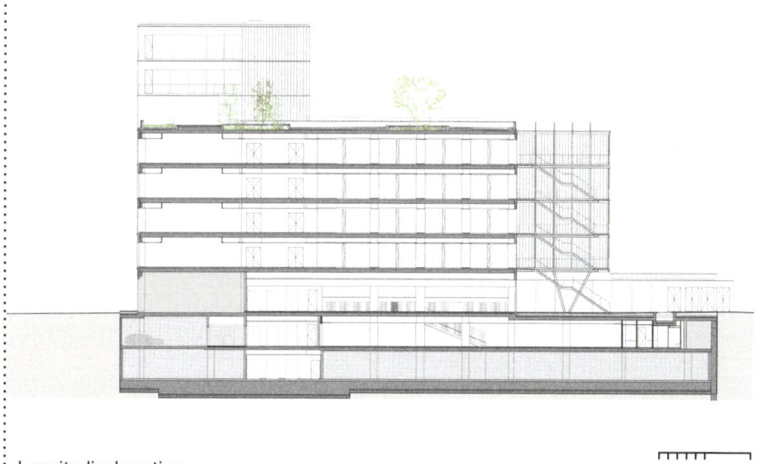

Longitudinal section

Building

On the fifth floor, the design incorporates a large 800 m² terrace, with a 300 m² porch at one end.

A system of visible installations that allow for open heights of 3.15 meters complements the open-plan office floors. The placement of the false ceiling through closed perimeter strips that hide bulky and noisy system components allows for clean, organized, light-filled spaces adapted to the needs of the technology companies in 22@.

The vertical garden in the entrance hall, with ample entry of natural light through a skylight, communicates with the basement through a singular staircase.

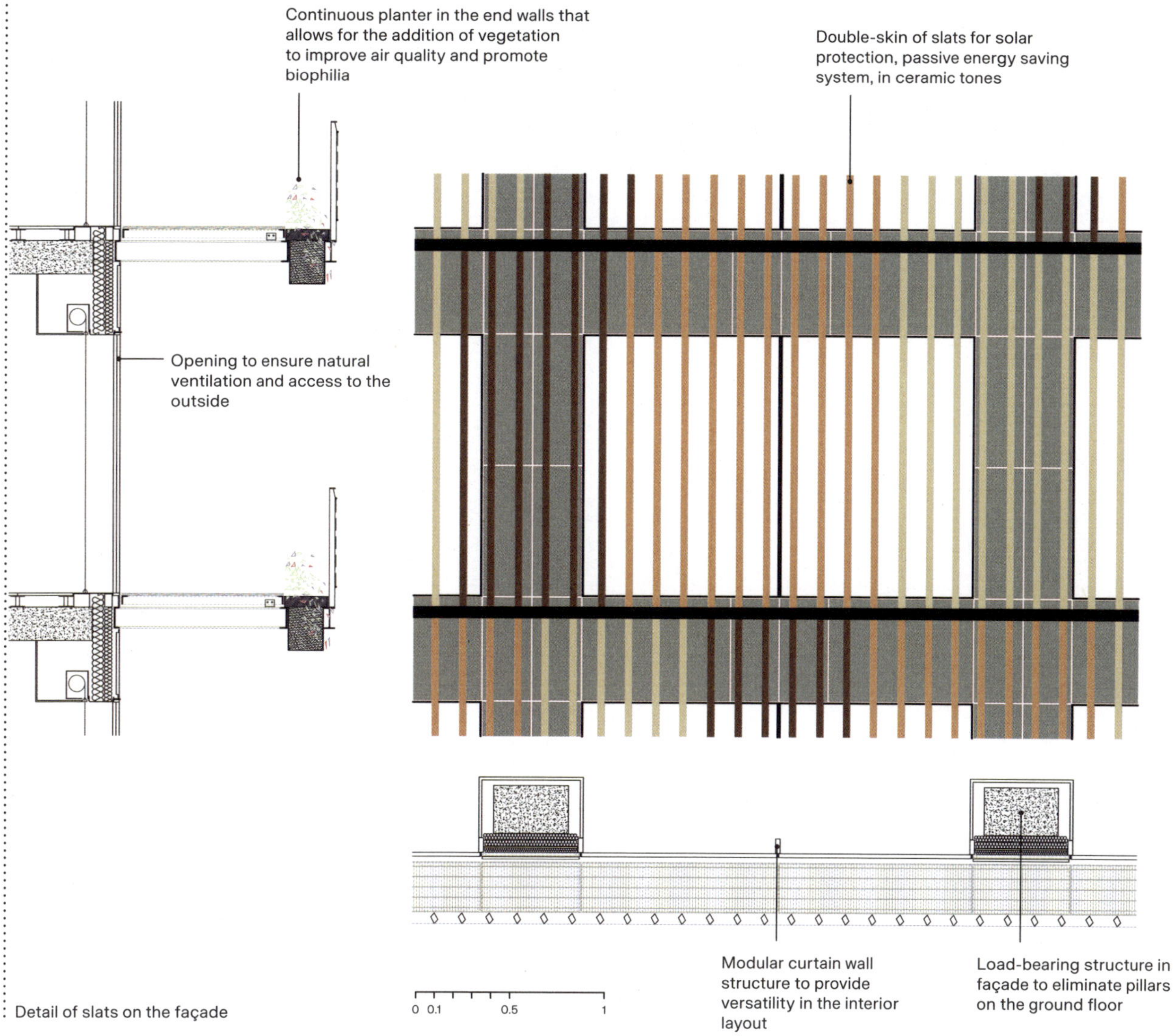

Detail of slats on the façade

Thanks to passive solar protection and an efficient air conditioning system, the building has the highest A energy rating, as well as LEED Gold V4 certification. This guarantees that the design meets environmental and reduced water consumption and energy demand criteria. It also guarantees the use of local materials manufactured according to sustainability criteria in the building's construction. The project provides a total of 305 m^2 of landscaped areas in the different open spaces and on rooftops.

Client Esecetanger / **Authors** Enric Batlle, Joan Roig, Albert Gil / **Team** Meritxell Moyà, Cristòfol Tauler, Manel Ventura, Diana Calicó, David Sobrino / **Collaborators** STATIC – Structural engineering, PGI Engineering – Facilities engineering, Envolventes Arquitectónicas ENAR – Façade consultancy / **Contractor** OHL – Obrascón Huarte Lain / **Gross floor area** 11,585 m^2, 1,222 m^2 (exterior)

Using Small-Scale Ecological Strategies

Small-scale ecological strategies are based on ingenuity and hope. They are the product of the conviction that we have to do new things to have better cities. They are the result of small contributions made at community level to support global sustainability. While they are not the solution to all the world's ecological problems, they can help chart a course that brings us closer to it.

Small-scale ecological strategies are a response by cities to environmental crises. As Martí Boada says, the culture of hope is growing in cities, and from this culture a model of survival has to emerge. These strategies are a local response to global problems. They are actions we can take close to where we live as an alternative to large-scale energy production, water supply, waste accumulation and food production operations.

Self-Sufficiency

Generating in the Neighborhood, Combining Energies

Producing Energy in the Public Space

100% Positive Energy Balance

Generating in the Neighborhood, Combining Energies

Concentrating production and consumption of resources in various places creates cities that extract resources from nature and are incompatible with a sustainable model. Within our urban fabrics we must develop energy production and resource alternatives with the aim of achieving emission-free cities. By reducing consumption and generating energy in our neighborhoods, we can ensure that urban areas produce the resources they need to be self-sufficient.

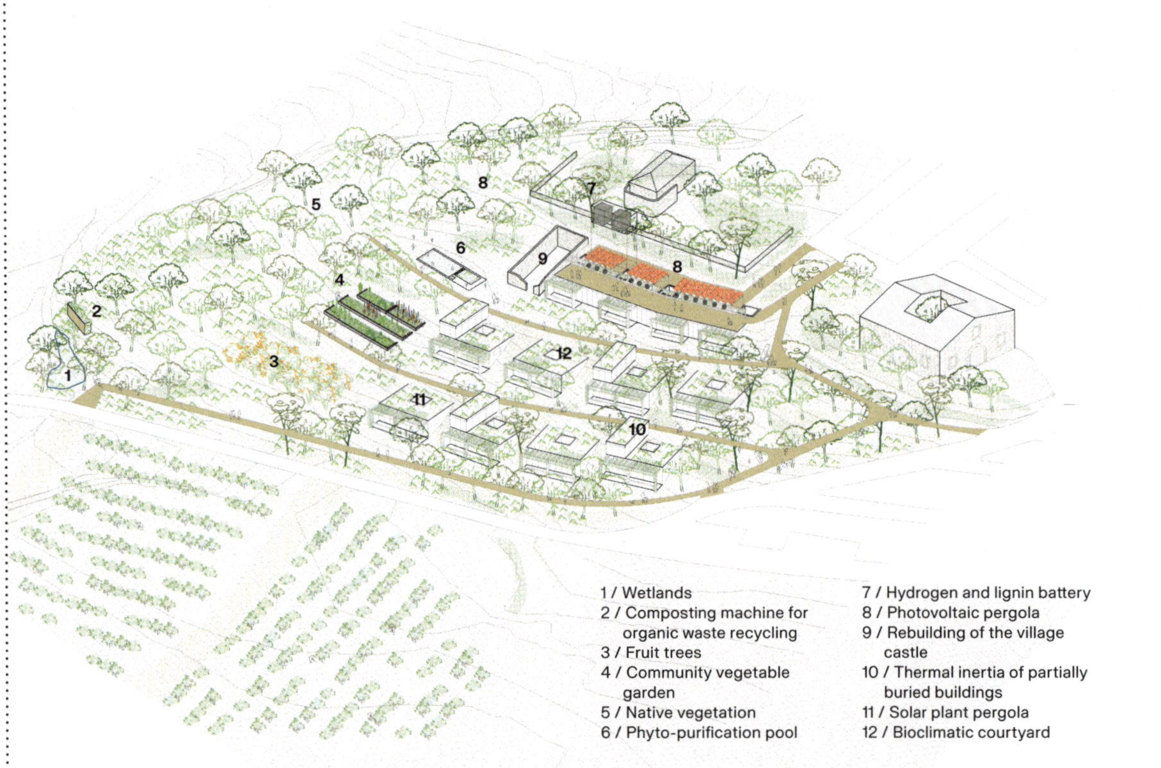

In new urban development projects of a certain size, we need to think about closing hydrologic, material and, particularly, energy cycles. In the case of **the Eco-neighborhood in Vilamaniscle**, a village in the Empordà region, passive housing systems and local energy production create a closed production and consumption cycle within the community.

Using all available energy resources, both natural and artificial

We need to harness renewable energy sources from natural sources (solar, wind, geothermic) for both the city and natural habitats, as well as energy generated by human activity such as residual heat and cold from industrial activities or gas from urban waste.

Networking to maximize joint system performance

System balance requires all elements to be connected in a network so that roofs that receive sunlight or industries that generate heat can be connected to offices that need electricity or homes that need heat, all managed in a professional, centralized way with continuous system updating.

Taking advantage of the synergies created by a combination of uses and the city's economy of scale

The combination of uses with various energy balances at separate times allows us to coordinate our efforts to balance the generation and consumption of energy in such a way that demand stays as constant as possible and can be met with the least environmental and economic cost.

Transformation of Blocks in Barcelona's 22@ District (Spain, 2006, 2008, 2019 and 2021): a sustainable recovery of a historical industrial fabric.

100% self-sufficiency = passive measures + centralized air-conditioning network + local production of energy

20,000 fewer tonnes of CO_2 emitted into the atmosphere

1°C and **2°C** below ambient temperature

The Poble Nou neighborhood, one of the largest centers of Spanish industrialization in the eighteen-hundreds, had entered a period of decline by the second half of the 20th century. Twenty years ago, the city government proposed a major urban renovation project involving 200 hectares that encourages the implementation of mainly tertiary uses –preferably connected to technological activities and the new economy– as well as a substantial amount of social housing, with the aim of transforming the neighborhood into the city's main economic and technological platform.

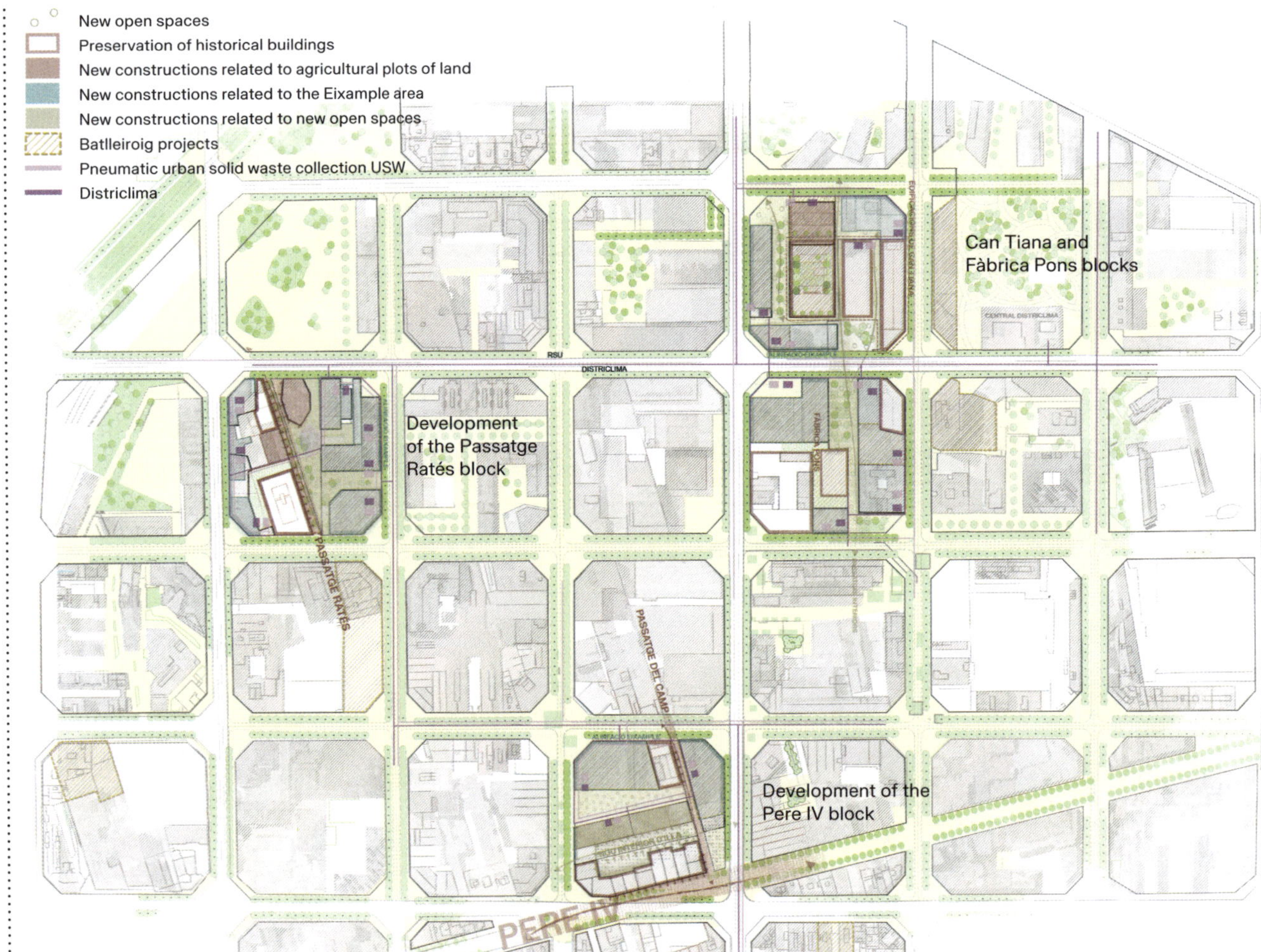

Plan of the 22@ district

The plan lays the groundwork for a gradual block-by-block transformation, taking advantage of the support provided by Cerdà's Eixample grid: each block is developed according to an urban improvement plan (PMU), the goal of which is to organize the new uses envisaged by adapting to the particularities of its preexisting physical elements and plots of land, preserving consolidated buildings, as far as possible reusing the industrial architectural heritage and highlighting the structure of passages running through the blocks.
The Cerdà grid also acts as a support in the neighborhood for a Special Infrastructure Plan, a network of highly competitive services: telecommunications, pneumatic waste collection and infrastructures committed to energy innovation whose focal point is the largest heating and cooling distribution network in Spain, the Districlima network that minimizes the consumption of fossil-based primary energies. It prioritizes energy efficiency, control and reduction of noise pollution, and the responsible management of natural resources.

This comprehensive strategy continues with the derived planning designs, block by block, where passive strategies such as the recovery of existing buildings, an urban approach mindful of sunlight and orographic winds, and the fashioning of skins for smart building, with solar protection and natural ventilation, serve to reduce demand significantly. Finally, the sustainable production of energy in the building itself completes this self-sufficiency.

Can Tiana and Fàbrica Pons blocks (2008): a development involving a public space associated with the recovery of industrial heritage
The two blocks, the focus of the overall operation, bear the names of the factories constructed there in the 19th and 20th centuries, Can Tiana and Fàbrica Pons, now Heritage Catalog protected sites. The factories were built in the middle of the block and, over time, adjacent auxiliary warehouses and higher buildings with different uses were added on the perimeter, maintaining the interior structure of service passages.
The urban development project consists of preserving the historical industrial facilities, maintaining the structure of the original passages and expanding with new green areas, creating alternative routes within the Cerdà district that connect through pedestrian spaces to Plaça de les Glóries.

Can Tiana and Fàbrica Pons block (2008)

The new office and social housing buildings are organized along the perimeter of the block, connected both to outside roads and internal passages, with different heights and a special volume to house the headquarters of the Telecommunications Market Commission, a public body.

Development of the Pere IV Block (2019): a transformation that is the result of respect for the original design

This block is the result of the superimposition of Cerdà's Eixample grid on the pre-existing design: Carrer Pere IV, the former city approach road to the north, that cuts through the perimeter of the block, and the agricultural division of land that, following the direction of this street, gave rise to the block's two irregularly preserved interior passages. The block transformation project, which maintains part of the existing building, ensures communication between these two overlapping logics: on the one hand, it recovers and reinforces the interior passages, which structure the open spaces and the municipal complex consisting of public housing and a facility located in the center of the block; and on the other, it aligns the new office buildings with Cerdá's grid, with a single volume and specific heights overlooking Carrer Badajoz, a street in the process of being transformed into a major tertiary axis.

Development of the Pere IV Block (2019)

Development of the Ratés Block (2006, 2021): a historical passageway transformed into a central public space

The Passatge Ratés Block, located at the western end of 22@, is distinctive for the passageway running through it with a consolidated front of low-rise housing, and an industrial building listed in the Heritage Catalog. The Urban Improvement Plan responsible for transforming the block has to specify the volumetric arrangement of the new uses while maintaining the existing building.

The proposal preserves the passageway and extends it through the transfer of public land for open spaces and facilities, thereby giving continuity to the public spaces planned for the adjacent block.

The passageway also divides the block into two types of construction. On one side are the buildings that will be preserved, requiring the others to conform to a complex geometry that respects the current land division to ensure a coherent layout adapted to existing party walls. On the other side, the building is freely arranged on a new plot of land, in three separate volumes with their respective designs and building phases.

Development of the Passatge Ratés Block (2006, 2021)

CAN TIANA AND FÀBRICA PONS BLOCK PMU. Client Grupo Castellví / **Authors** Enric Batlle, Joan Roig / **Team** Goretti Guillén, Abel Porcar / **Collaborators** José Soria – Lawyer / **Project area** 20,925 m², 62,774 m² (gross floor area) // **PERE IV BLOCK PMU. Client** Tramway Capital / **Authors** Enric Batlle, Joan Roig, with Gina Barcelona Architects AIE, Elvira Fernández (Pareja i Associats, Lawyers) / **Team** Abel Porcar, Helena Valls / **Collaborators** Intra – Mobility studio / **Project area** 9,120 m², 27,315 m² (gross floor area) // **PASSATGE RATÉS BLOCK PMU. Client** Brilten / **Authors** Enric Batlle Durany, Joan Roig / **Team** Abel Porcar, Albert Gil, Ronda Quetglas / **Project area** 10,483 m², 31,448 m² (gross floor area).

Producing Energy in Public Space

Viewing the landscape as a productive element means designing public spaces that do not depend on services outside the project. These spaces will be self-sufficient throughout their lives for their intended use. Designs must incorporate the natural evolution of the landscape and its vegetation, as well as the energy supply for its installations and equipment. We can take advantage of the public space to collect rainwater, urban elements to produce energy, and vegetation to generate composting material and perhaps biomass as well.

The landscape around **the Parc des Expositions MEETT in Toulouse** proportionately combines the spaces devoted to vegetation, rainwater management and photovoltaic energy consumption.

Analyzing demands and seeking self-sufficiency

We need to design public spaces to facilitate the collection of rainwater for later use to irrigate vegetation. We need to plant species with low water requirements and use organic waste to generate composted material for our public space.

Harnessing public space to produce energy

We should view public space as a productive space in which to place panels, urban wind turbines and other technological developments that help us move closer to achieving self-sufficiency. We can equip urban elements such as pergolas and buildings with photovoltaic panels that generate the electricity we expect to consume and even create a positive energy balance.

Characterizing public spaces according to energy needs

We have to design public spaces based on positive discrimination of lighting levels according to expected uses. We need efficient street lighting systems controlled by photosensitive sensors for optimal on-off management.

Green decking of the Ronda de Dalt Beltway in Barcelona (Spain, 2018-2020): a productive wooden pergola transforms a public space into a biodiverse climate refuge that connects two neighboring districts.

43.25 kWp of energy generated for self-consumption by local facilities, equivalent to the consumption of 27 homes

9,021 m² of land reclaimed from cars Phases 1 and 2

25% shade in relation to the promenade Phases 1 and 2

Barcelona's emblematic beltway – Les Rondes – was designed in the 1960s and developed thirty years later thanks to the impetus of the 1992 Barcelona Olympic Games. The construction work involved an unprecedented urban transformation in the city.
While the Ronda de Dalt was a road infrastructure success story, its shortcomings in terms of public space quickly became apparent and plans to make up for this deficiency were soon in the works. First, it was necessary to recover the continuity of neighborhoods that had been divided since the late 1980s.

Thirty years later, social activity in front of Vall d'Hebron Market, between the neighborhoods of La Teixonera and Sant Genís dels Agudells, where it seemed to have disappeared almost completely, has returned. A new mobility paradigm has shifted the central focus away from cars, reclaiming the infrastructure as an active part of the system of open spaces in our city that prioritizes walking, interaction between neighbors and traffic calming.

A linear garden
A strip of variable-width vegetation on the mountain side becomes an ecological connector between Collserola Park and the different green spaces around the beltway, extending the Mediterranean forest landscape of the mountain range over the new civic artery.

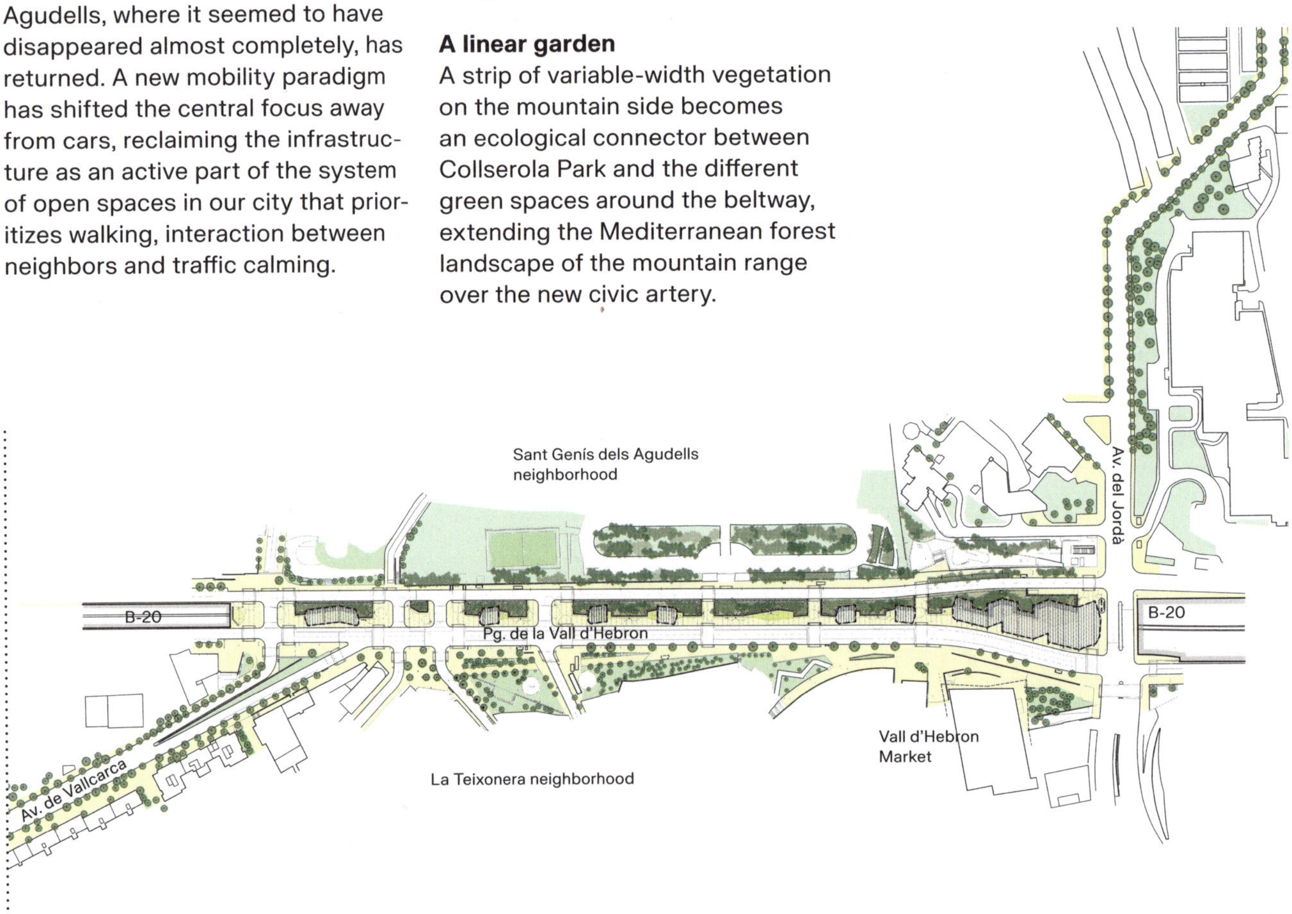

General plan

Section

The new promenade, conceived as a large multipurpose plaza due to its size, is capable of hosting open-air markets to complement the regular Vall d'Hebron market. The spacious plaza is shaded by a large productive 1.6-hectare photovoltaic pergola, the city's largest.

The dramatic asymmetry of the surroundings makes the enhancement of this section possible. A linear garden of native species will serve to renaturalize this part of the city while the large productive pergola maximizes the presence of vegetation that will gradually colonize it, providing shade to the plaza and promenade, and making this new social focal point a quality public space.

The main challenge in designing the new open space for the city, gained by covering the road infrastructure with a large concrete slab, was to provide this new space with the best climatic conditions for use and to contribute to Barcelona's commitment to reversing the climate emergency we are now facing.

Productive pergolas
For the pergolas, we chose wood, a natural, sustainable and renewable material with excellent technical qualities for use in virtually any type of construction. These structures have also represented an opportunity to turn the city landscape into a productive terrain by placing photovoltaic panels on the largest pergola. With an installed capacity of 43 kW, the pergola will provide clean energy for nearby public facilities.

This design addresses the climate emergency in which we find ourselves with a new public space that unites the city, cultivating life, using natural materials and producing renewable energy.

Client BIMSA – Barcelona d'Infraestructures Municipals / **Authors** Enric Batlle, Joan Roig, Iván Sánchez, Joan Batlle, Lola Simón, Maria Blanco / **Team** Marina Santos, Mercè Lorente, Diana Calicó, Elisabeth Torregrosa, Yago Cavaller, Dolors Feu / **Collaborators** BIS structures – Structural engineering / **Contractors** VIAS, Voracys / **Project area** 7,630 m²

100% Positive Energy Balance

We can achieve energy self-sufficiency by producing 100% of our needs in the building itself with renewable energies. The energy efficiency of the systems we implement, the passive characteristics we can incorporate into our building designs and currently available renewable energies technology allow us to think about zero net energy buildings and even buildings that produce more energy than they need and that can form part of nearby micro networks to help reduce the consumption of more distant systems.

The Headquarters of Pull&Bear in Narón (A Coruña), uses surplus heat energy from adjacent industrial units to help with air conditioning throughout the office building.

Achieving a positive energy balance

An EB+ building (positive energy balance) produces more energy than it consumes. It is not a matter of its being disconnected from the network since it consumes power from the grid. Rather the energy it exports to the grid is equal to or greater than what it consumes.

Using the most appropriate renewable energy sources

The geographic location and climate determine which sources are the most appropriate for each project: photovoltaic panels for climates with lots of hours of solar exposure, wind turbines for intensely windy climates, geothermal energy in areas with subsoils with water tables, biomass for areas with surplus timber or managed forests.

Managing surplus energy according to proximity criteria

We need to develop harvest map analyses that detect possible energy complementarities in nearby buildings so that the surpluses of one can serve the needs of another.

Inditex Headquarters in Arteixo, A Coruña (Spain, 2015-ongoing): a highly sustainable corporate campus where large buildings are laid out around abundant green spaces.

3,312 photovoltaic panels

1,711 MWh annual production

6,624 m² photovoltaic field

LEED GOLD Certification **Central services**

LEED PLATINUM Certification **Zara.com**

The transformation of the Inditex industrial complex in Arteixo into a more people-friendly, green, integrated and healthy premises, carried out in different stages over the years, is currently in its final phases. The first phase, which included the construction of three buildings to house a dining hall, an auditorium, a gym and technology offices, laid the groundwork for the overall strategy: the complex was made more spacious by interspersing generous outdoor spaces between the new buildings, connected to each other and to the original buildings through a cloister-like system of outdoor walkways.

In 2020 projects began on the opposite side of the original buildings. A new office and photo studio building for Zara.com was implemented tangentially to the extension of one of the walkway axes, directly connecting this new building to the rest of the complex.

Inditex headquarters service buildings: mixed building with auditorium, cafeteria, gymnasium and offices
2015-2018

Future building for Zara Comercial
2022-under construction

Woods
2021-2022

New cafeteria for Zara Comercial
2022-in project phase

We are currently working on projects to complete the green perimeter ring and two new buildings: a second dining hall and a new 170,000 m^2 five-story office building for Zara's four departments, located behind the Zara.com building and parallel to the existing ones. The new building, with 233 meters of façade, is characterized by its horizontality, formal simplicity, sustainability and ample open work spaces designed to enhance a horizontal work dynamic among the different departments.

An organized, green, connected and sustainable complex
The final result will be a well-organized group of buildings, with different types of outdoor spaces, contributing to the creation of healthy areas arranged around highly sustainable buildings that will exchange energy with each other through their own network. An energy ring will connect the buildings to each other and to photovoltaic generation points for streamlined management and improved self-sufficiency.

The complex will generate a total of 1,711.1 MWh/year, thanks to the installation, first, of a large pergola with 6,000 m^2 of photovoltaic plates (2,768 panels) that cover the 638-vehicle parking area, providing the space with shade; and, secondly, a 1,088 m^2 pergola on the roof of the Zara.com building (544 panels), which will protect the roof from the solar incidence angle effect. In total, it will cover 45% of the energy that the zara.com building consumes.

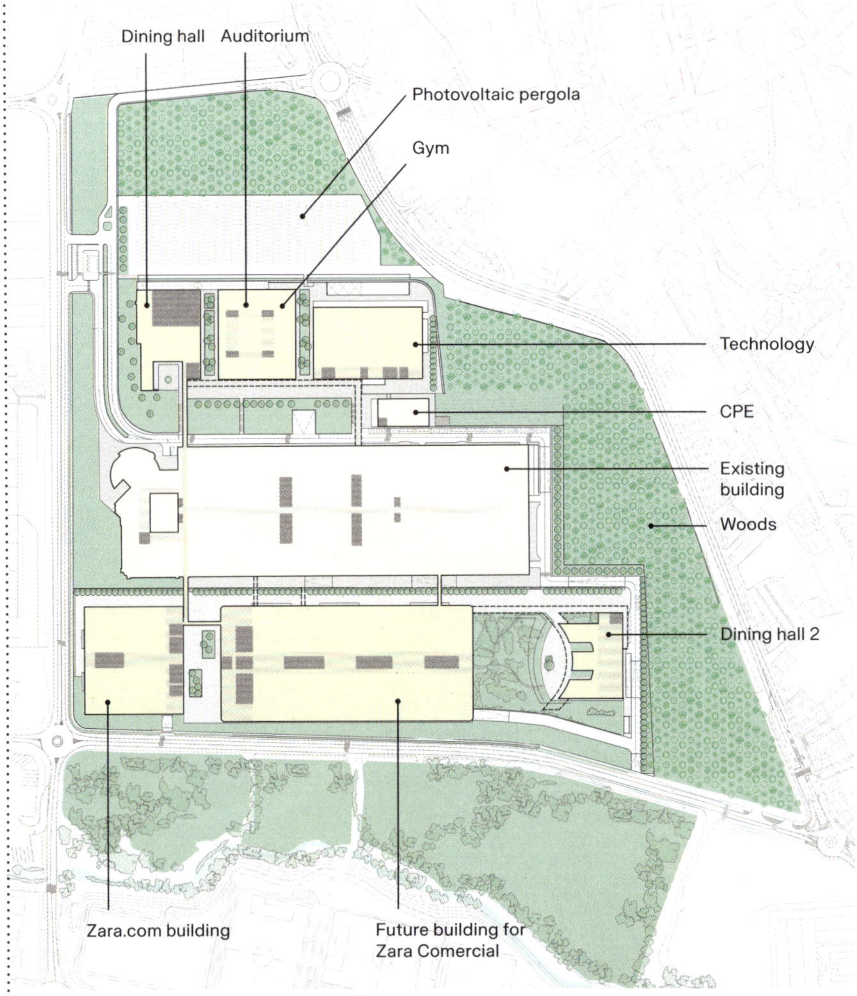

General plan of the complex

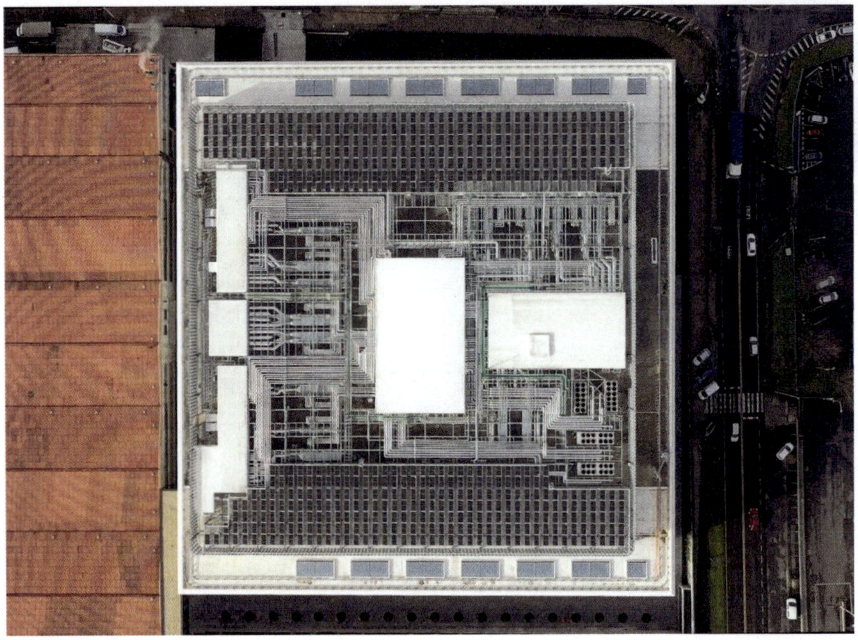

Zara.com roof

The perimetral green ring that provides the complex with wooded surrounding areas, thereby creating a smoother transition within the industrial setting in which it is located, is being developed in different work phases. As a result, some of the CO_2 emissions generated by the construction process have been offset. In phase 2, a 480-tree forest was created on a 33,500 m^2 green carpet that increases air quality and enhances rainwater infiltration on the existing site. In the current phase, the adjacent plots located to the south and west will be incorporated into this perimeter ring.

Zara.com building

Pergola seen through the forest

Service buildings from left to right: auditorium, gym and offices

A new walkway that organizes the complex

The first 70,000 m^2 complex was built overlooking a large plaza located at an elevation of 31 meters, where the company's main activities take place. The three new buildings have different volumes and are connected by means of a perimeter porch that externally links the different entrances and ends in two walkways to the buildings, over the interior roads. As a result, all the structures of the complex are connected in a pedestrian-friendly way by interior and covered spaces.

Walkway connecting the dining hall to existing buildings

The dining hall building at Inditex's headquarters has a prime location and a striking appearance compared to the rest of the building. Its structure is integrated into the building's window system. It is characterized by its transparency and excellent relationship to outdoor spaces, giving rise to three distinct dining areas with a capacity to accommodate 700 people.

Dining hall. Free flow

The other two volumes stand out for the horizontal-slat façades that give them a uniform appearance and provide a high degree of solar protection, thereby minimizing energy consumption. A large foyer open to natural light provides access to the auditorium and gym, while the more private and social areas are fitted with hardwood and materials with high acoustic absorption.

Collaborative working areas

Zara.com building
The new 63,184 m^2 Zara.com building with five 93 × 97 m floors is characterized by a structural system integrated into the façade that defines the building's image, a skeleton of 780 large windows through which natural light floods into large workspaces with structural spans of 20 meters and clear heights of 4.90 meters on the office floors and 6.50 meters in the photographic studios. Incorporating the bulk of the pillars into the outer skin substantially improves thermal insulation and dramatically reduces its energy demand. Repeating the modules creates an elegant, dignified exterior characterized by verticality.

Connection walkway between existing buildings and the Zara.com building

The Zara.com building is connected to the existing buildings by means of a walkway located on level 3 that completes a large internal connection axis between the new, the existing, and the first-phase buildings in an accessible, intuitive way.

To improve internal communication, the five floors are connected by wide stairways, promoting healthy mobility in the building. The main stairway is built in a large vertical atrium, 32 meters high, flooded with daylight. Creating comfortable stairways with natural lighting and situating them in visible places encourages their use and, therefore, people's wellbeing, reducing the use of elevators and their associated energy consumption.

Large clearance heights of up to six meters on some floors allow for a lower concentration of people per m^3. This, along with air renewal systems that filter harmful particles, guarantees high-quality interior air in a very low-density environment.

Zara.com photography sets

INDITEX HEADQUARTERS SERVICE BUILDINGS. Client Inditex / **Authors** Enric Batlle, Joan Roig, Albert Gil / **Team** Esther Díaz, Mariano Aguado, Beatriz Barral, Albert Solé, Adelardo Domingo Ruiz, David Sesé, Segundo Rodríguez, Maria Duran, Josu Acebrón, Diana Calicó, Ana Anido, Dolors Feu / **Collaborators** Keyplan – Engineering, DM Ingeniería, Estudis ESL – Facilities Engineering, Arau Acústica – Acoustics Engineering, Instra Ingenieros – LEED Consulting, Antonio Reboreda Ingeniería – Structural Engineering, Prosegur – Security installations / **Contractor** Goa Invest / **Gross floor area** 78,845 m² // **BUILDING FOR ZARA.COM. Client** Inditex / **Authors** Enric Batlle, Joan Roig, Albert Gil / **Team** Beatriz Barral, Mariano Aguado, María Méndez, Paula Ortiz, Claudia Márquez / **Collaborators** Keyplan – Facilities Engineering, Daie – Structural Engineering, Enplano – LEED Consultancy, Galaicontrol – Quality Control, SGS - Commissioning / **Contractor** Goa Invest / **Gross floor area** 63,184 m²

Putting People at the Center

Richard Sennett argues in *Building and Living* that contemporary urbanism has become a fractured discipline, divided between knowledge about building and knowledge about living. To overcome this rift, perhaps we should return to the origins of the discipline, as in the time of modernism, when the two were inseparable. We need to build a city that is open, inclusive, healthy, porous, habitable and full of interactions and infiltrations instead of sticking to the closed-border paradigm that dominates the modern city, where urban habitats are cut up into parts separated by traffic flows, and different use and income areas are isolated.

We have to put people back at the center of the discussion of how we design our cities, public spaces and buildings, how we adapt to the climate emergency and the challenges arising from the pandemic and, ultimately, how we “re-merge” with nature.

LaMercedes Ecodistrict, Barcelona, 2021

Health and Well-Being

An Inclusive City

Healthy Public Spaces

WELL Buildings, the User at the Center

An Inclusive City

The necessary environmental revolution that coincides with the social revolution we are experiencing at the outset of this century must impact the way we design our cities. Increased awareness of the inequalities resulting from economic status, gender, functional diversity, racialization and age also call for reflection on the role that urban development can play in mitigating these inequalities. The city is simultaneously the reflection and foundation of our daily activities: its form and the physical organization of its different uses influence the ease or difficulty we experience when carrying them out. For this reason, the design of a city must consider the vital processes of the different groups that inhabit it.

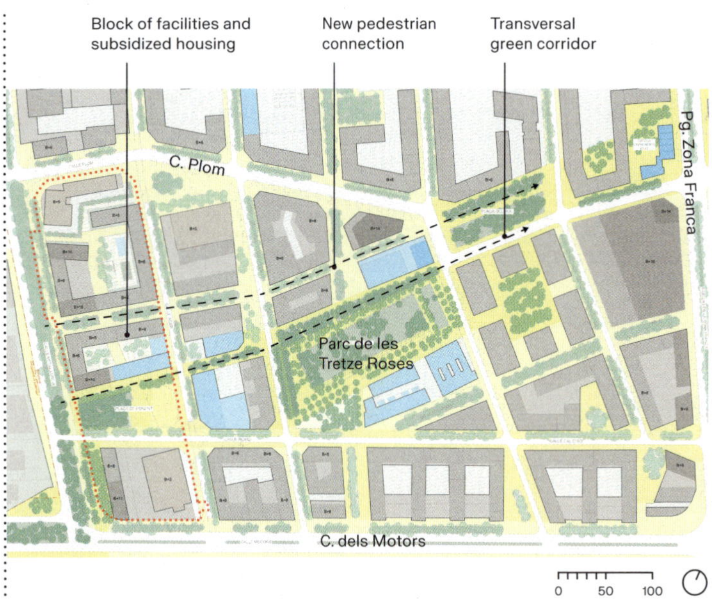

The Urban Project in the Marina del Prat Vermell District in Barcelona transforms an old industrial fabric by implementing a combination of tertiary uses (existing and new) and free-market and protected residential developments. High-intensity use of public spaces complements the social and functional mix needed to create an inclusive neighborhood, thanks to a mosaic of green areas and facilities. This form of urban design promotes use by the community as a whole and the inclusion that strengthens the sense of community belonging.

Conceiving of the city as an integrating mechanism

We need to design our cities with respect for the different existing urban and social fabrics, articulating them while establishing elements of continuity and cohesion between them. We need to foster the variety and flexibility of residential types to promote a mix of socio-economic statuses and avoid isolation and segregation. We also need to integrate memory of place, keeping in mind the important buildings and elements that represent a collective historical memory.

Making everyday life easier

Our goal should be to move toward a decentralized, compact, dense model with a combination of uses and types: a city with short distances that facilitates the multiple tasks of daily life. This city model should include coherent planning of buildings and public spaces to promote pedestrian accessibility, mobility and safety.

Promoting community life

Our cities need to be conducive to social relations and exchanges, temporal appropriation, intermediate spaces for neighborhood coexistence, meeting and play, and local relationships. The appropriation of space by a living social fabric fosters community care, socialization of life-supporting tasks and informal control. We have to encourage symbolic spaces endowed with a special meaning for a group of individuals to contribute to creating a collective identity and facilitate their interaction.

LaMercedes Ecodistrict in Barcelona
(Spain, 2019-ongoing): a mixed, compact and inclusive neighborhood. A place to work, play and live in an industrial and cultural complex prepared to address climate emergency.

100% of public space for pedestrians

60% permeable surface in open public spaces

+ 600 different tree species, **200** of which are fruit trees

The laMercedes project seeks the overall transformation of the site of the former Mercedes-Benz factory, which during the 1950s was the headquarters of the public company ENMASA and since its closure in 2007, more than a decade ago, has become a self-contained, inactive space.
The historical and heritage value of this nine-hectare site, classified as an industrial zone, lies both in the material value of its buildings, with a rationalist architecture of unmistakably Nordic inspiration, the work of the architect Robert Terradas, (the result of successive projects from the 1950s to 1970s), and in its intangible value, testimony to the neighborhood's industrial past, with which many of its residents are closely connected.

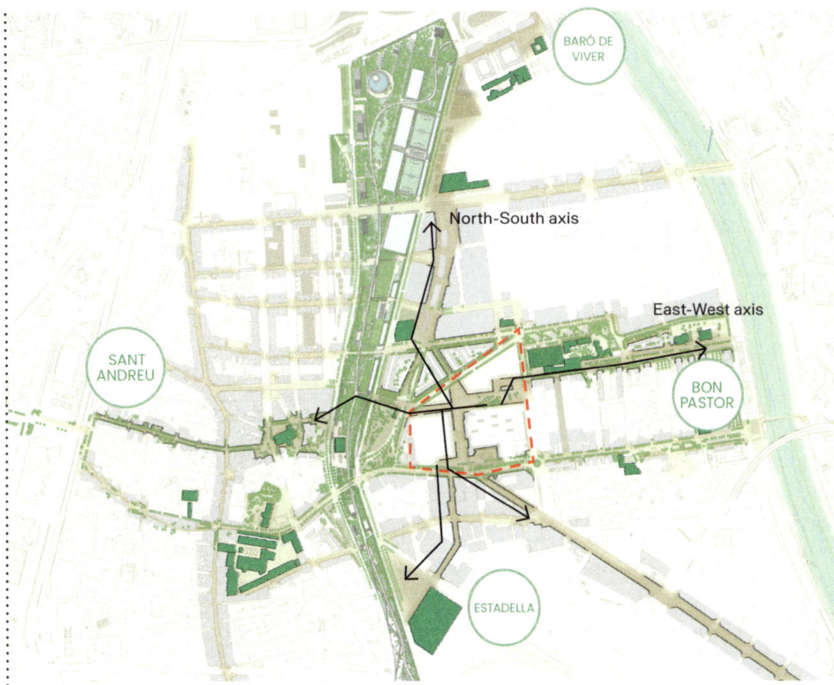

Civic activity layout

Stitching the city together
LaMercedes is located between two large metropolitan parks, one on the banks of the Besòs River and the future Rec Comtal park over the railroad tracks, which connect the city ecologically and socially to the natural areas of the Marina and Collserola hill range and the metropolitan coastline. The old industrial complex blocks connections between residential centres with highly differentiated social and economic characteristics. The proposal transforms it into a crossing point of two new urban axes: through the east-west axis the historical center of Sant Andreu will be connected to the Bon Pastor neighborhood and the future Rec Comtal park and Besòs River park, while the north-south axis will connect the Maquinista shopping area and the Baró de Viver neighborhood to Rec Comtal park and the future developments within Torrent de l'Estadella industrial park.

This is an urban development that recognizes the architectural history and social significance of the place based on the reuse of different existing buildings. The management and administration office, including the gardens and the restoration of the original lake, will be preserved as a gateway to the new neighborhood. Other elements that will be preserved around the square are the paint shop next to Carrer de la Ciutat d'Asunción and the high-rise workshop and office building from the 1970s. The entrance volume of the 1970s workshops at the rear of the site will also be maintained.

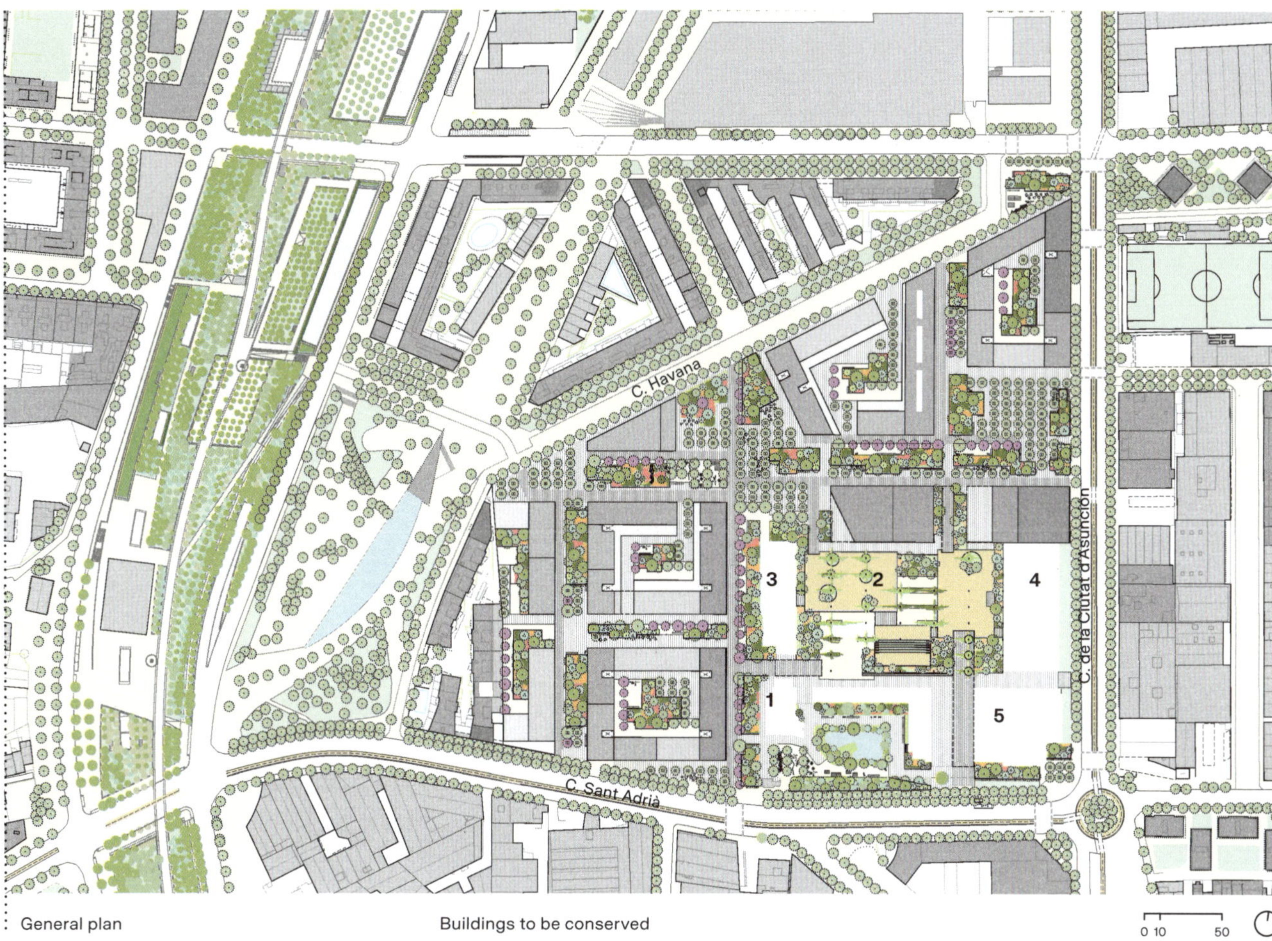

General plan

Buildings to be conserved

1/ Office building (1951)
2/ Production plant (1951)
3/ Factory (1951)
4/ Paint workshop (1960)
5/ 1970s workshops

A mixed-use neighborhood

The implementation of activities related to education and innovation in the historical buildings is complemented by new constructions for economic activities 4.0 (creative industries and knowledge and circular economy sectors). All of them are organized around a public plaza of over a hectare that recovers the structure of the historical factory building to host civic activities. Maintaining the character of the site and building this symbolic space contribute to creating group identity and a sense of belonging. In the rest of this sector, residential uses will be concentrated in several blocks of social housing – 40% of homes in the sector – and free-market housing organized around the main axes of the area. The transformation of this large and currently impermeable area into city is seen as a unique opportunity to work from the planning stage, that is, from its genesis, according to the premises of inclusive urbanism. The mix of uses and building types that characterizes the short-distance city facilitates management of everyday tasks and fosters the accessibility, mobility and safety of people who travel by foot.

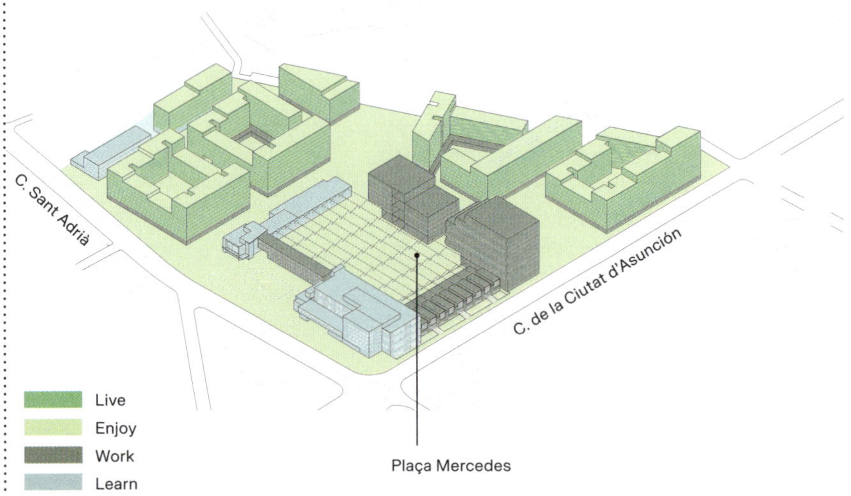

1 / Live
Housing and local businesses 116,000 m²
A new flexible, versatile and non-hierarchical housing model with habitable outdoor spaces

2 / Enjoy
Green axes and squares with abundant vegetation: 47,100 m^2
Plaça Mercedes: 10,000 m^2

3 / Work
The industrial buildings are reused to create new office buildings

4 / Learn
Private and public cultural and educational facilities recycle the site's main buildings

A soft mobility neighborhood with areas for new mobilities

The commitment to a neighborhood designed for sustainable mobility leads to a traffic-calmed permeable public space, with exclusive circulation for pedestrians and bicycles, and an efficient perimeter connection to the subway and commuter network that links the neighborhood with the entire metropolitan area.

A subterranean circulation system with entrances from the perimeter streets, in addition to distributing vehicles to the parking lots, will also accommodate last-mile distribution logistical spaces, loading and unloading operations for neighborhood businesses, and spaces for recharging electric and shared vehicles. These underground spaces will also house efficient service networks that structure the development as a circular system.

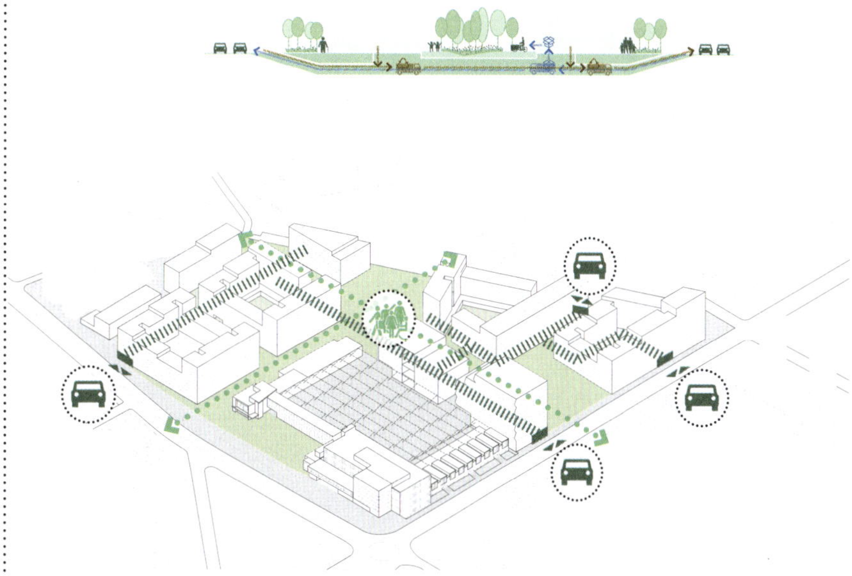

Rolling permeability + pedestrian permeability

Main axes
LaMercedes internal routes, integrated into nature, produce extensions, crossings, rest points, and play and leisure areas at different scales.

Green infrastructure.
Nature at the center of the design
The project seeks to expand green areas and continue to the urban axis of La Sagrera park to create a large green carpet on which to develop the new eco-city. Some 5.4 hectares of open space is organized around two main axes. This creates a hierarchy of streets and squares, routes and activity zones that will serve as garden-streets, with different natural environments, themed gardens and different types of vegetation. The street section will be variable, accommodating different uses in relation to the greenery and the first floors, enhancing biodiversity, green productivity and ecological connectivity.

Type A:
Maximize activity
Maximum section with expansion of promenade and maximum gravel width

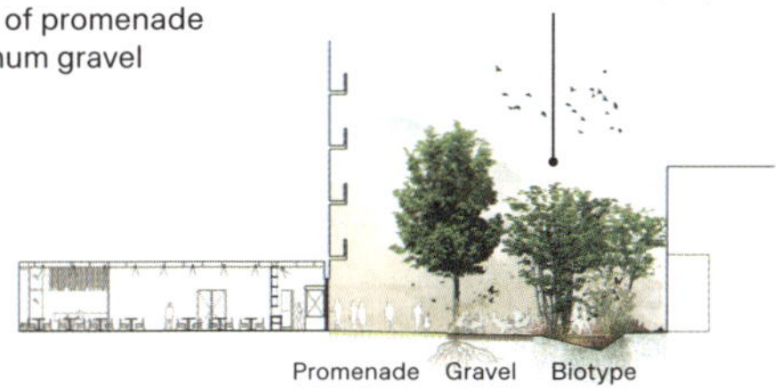

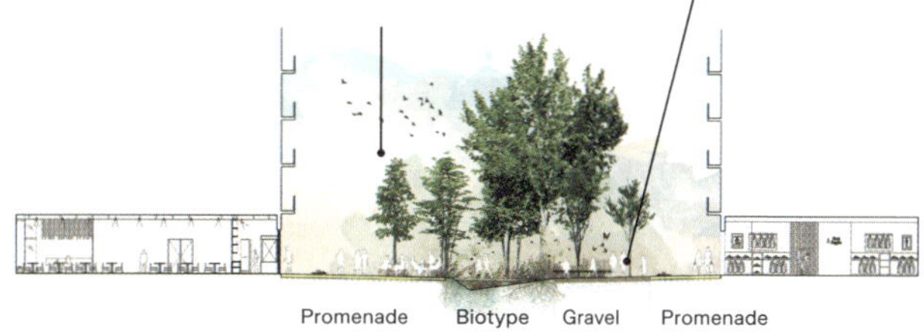

Type B:
Maximize biodiversity
Maximum breadth of biotype and water infiltration low point

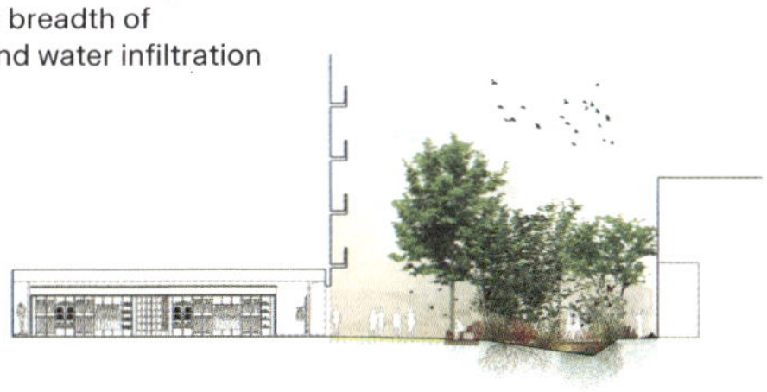

Sections. North-south axis. Main civic axis

Sections. East-west axis. Commercial axis

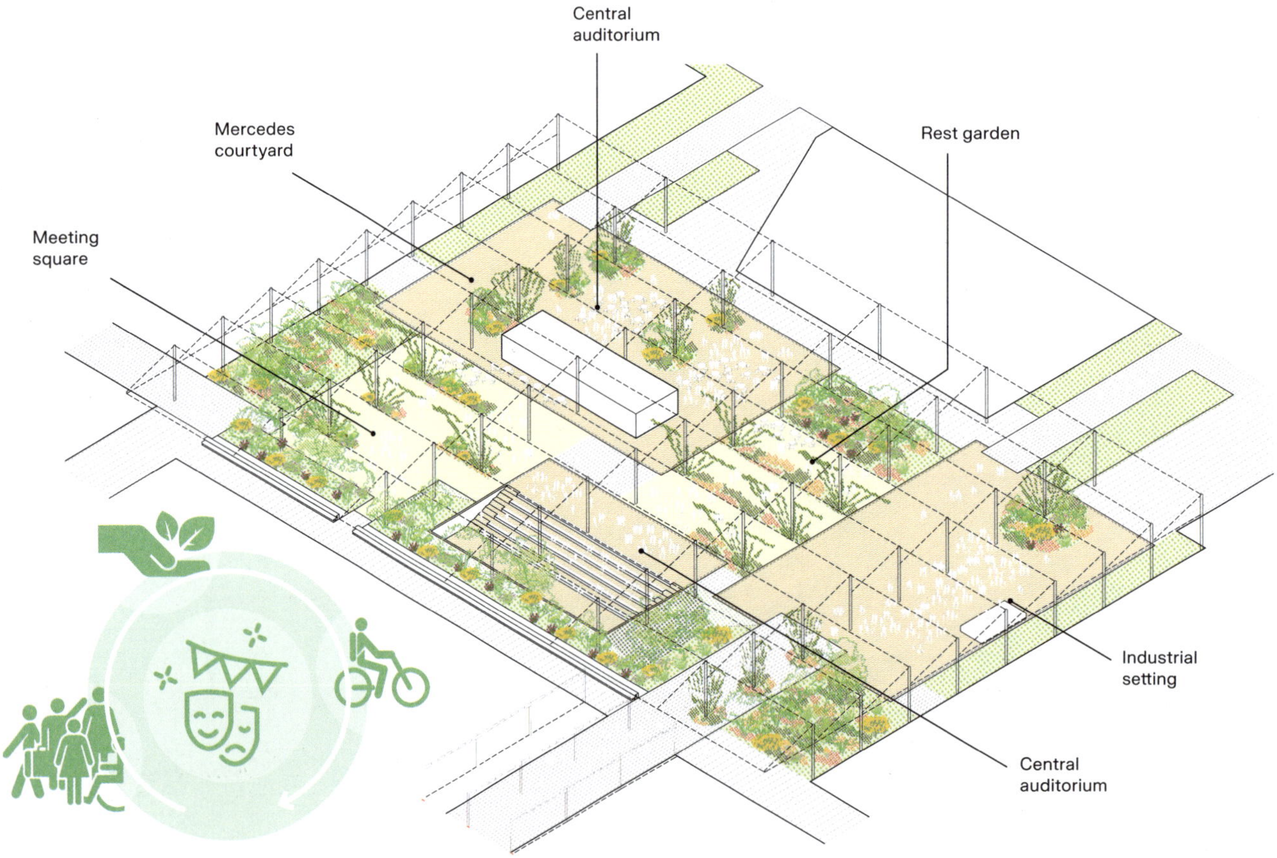

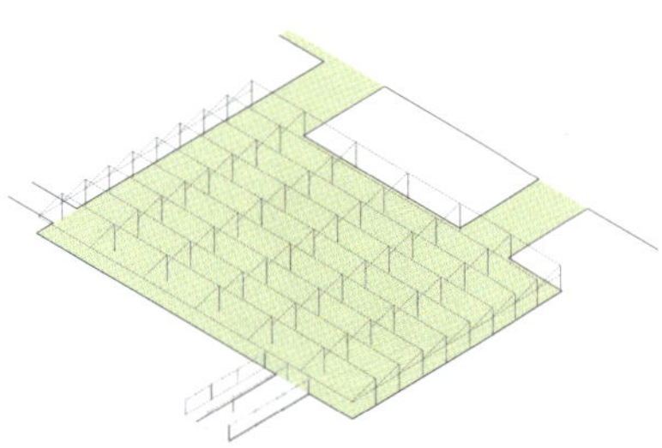

Green canopy

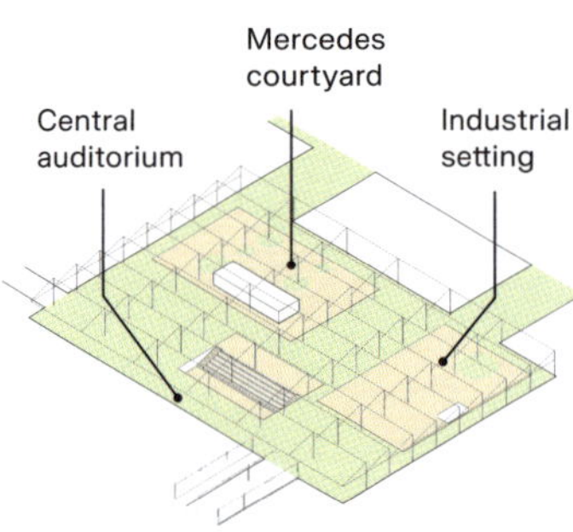

Main uses

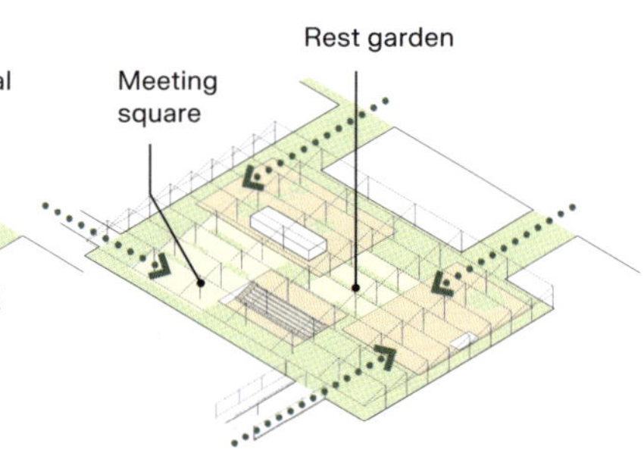

Interstitial spaces

Mercedes Square
A multipurpose permeable space for different activities that reuses the old factory building dating back to 1951. This is the central agora of laMercedes, a focal point and meeting place for the eco-village. A large canopy, a continuous and extensive green cloak, is proposed not only at ground level but also over the existing metal structure.

A new central open space of more than a hectare recovers the building's industrial structure to house civic activities.

Main space before work

Client Conren Tramway / **Authors** Enric Batlle, Joan Roig, Abel Porcar / **Team** Mercedes Blay, Helena Valls, Josep Maria Navarrete, Diana Martínez, Helena Salvadó, Albert Gil, Josep Batlle, Joan Batlle / **Collaborators** Societat orgànica – Engineering, Aiguasol – Facilities engineering, Urbaning (Ideem Innova), Garrigues – Lawyers, Folch – Graphic Design / **Project area** 90,560 m², 200,667 m² (gross floor area).

City and Territory

Healthy Public Spaces

Biophilic cities are cities that contain an abundance of nature. They are cities concerned with protecting and restoring the natural surroundings, striving to foster deep connections with the natural world so that we can enjoy it indefinitely. Nature is not an option; it is essential for living a happy, healthy and meaningful life.

The Renaturalization Project of the Qurtubah Neighborhood in Riyadh (Saudi Arabia) involves a commitment to comprehensively transform an entire neighborhood by planting more than 23,000 trees. The proposal seeks to encourage use of public spaces by creating a grid network of shaded paths that connect the main public spaces and facilities in the neighborhood.

Inhabiting public space

We need to design our cities to be open, inclusive, healthy, porous, inhabitable and full of interactions and infiltrations that help us to overcome the modern city model of segregated spaces divided by traffic flows and classified according to specific uses or different incomes.

Promoting biophilic cities

We need to put nature at the center of our decisions, given that it is one of the best strategies for improving people's health and well-being. Maximizing the presence of trees and all kinds of vegetation in our urban surroundings influences the health of our citizens.

Designing inclusive spaces

We must define our public spaces from a gender and accessibility perspective to guarantee the safety and comfort of all citizens, from childhood to old age, including all social groups, particularly the most vulnerable. Places to rest with good orientation surrounded by vegetation should be incorporated into open spaces to make them healthy, pollution-free areas ideal for walking that provide serenity and silence, and encourage social interaction.

Scenic Route in the Old Gypsum Mines of Igualada, Barcelona (Spain, 2016-2018): a new healthy route that recovers a natural space using water-based technical solutions and includes high-tech building solutions for the paths.

850 m of path activating a 20,000 m² area

100% rainwater infiltration

4-m-wide, accessible, inclusive and healthy path

The proposal is part of the general Igualada Green Belt project, which seeks to create a network of pedestrian and cycle routes in the form of a green belt around the perimeter of the city. The natural space through which the path runs have become very run-down, the site of intense gypsum mining activity in its subsoil (Guixeres del Claret and Pelfort) and later used as a municipal waste transfer station.
The built section, about 800 m, is located to the north of the city. It consists of three distinct strips that adapt to the different conditions along the route, addressing the problems of the site while also developing its potential.

The project has two main objectives:

1 / Connectivity: The continuity of this overall action at the city scale is achieved by means of a high-tech path surface with optimum performance in terms of accessibility, visibility and durability.

2 / Landscape and biodiversity recovery: the necessary elements are provided to foster a new dynamic on the site that progressively improves its environmental conditions using water.

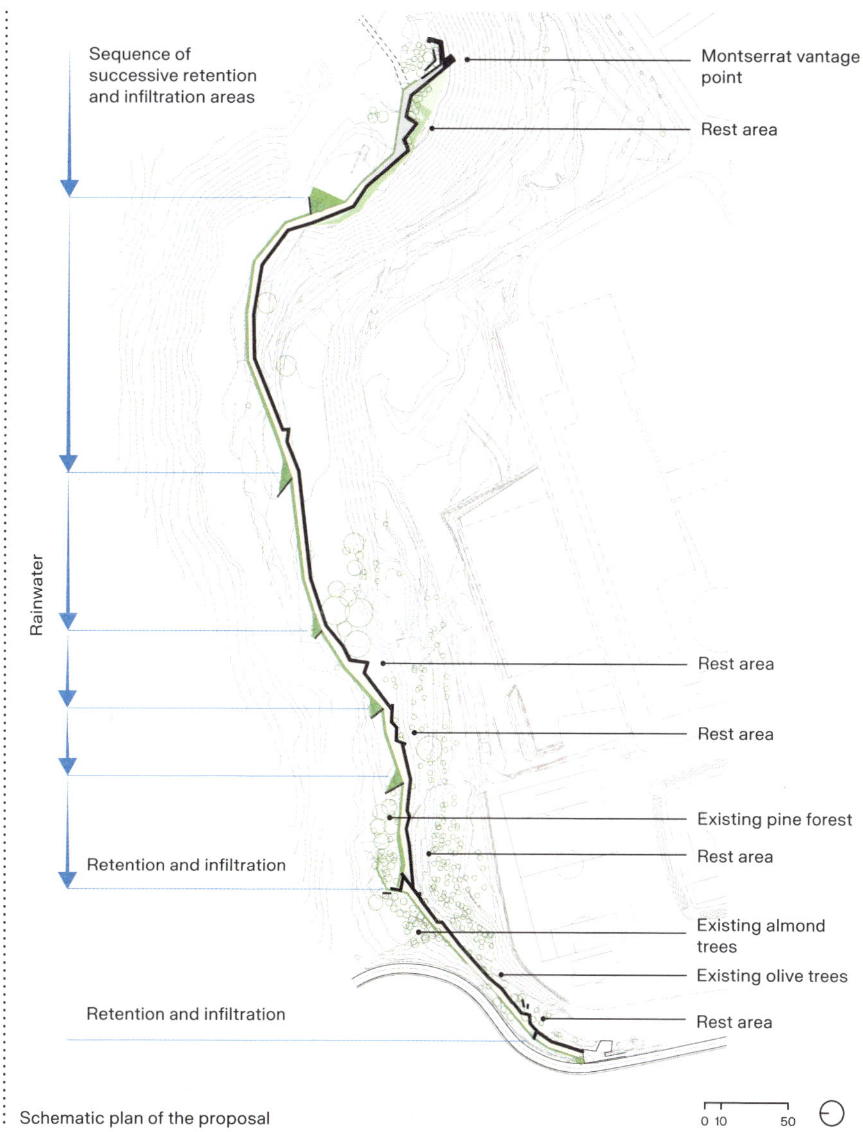

Schematic plan of the proposal

Water is a precious resource. With this in mind, the proposal entails a new drainage system to control surface water through planted ditches and the creation of a series of retention areas on the hill side of the route to allow rainwater infiltration. This, in turn, generates a safety strip to retain any small landslides.

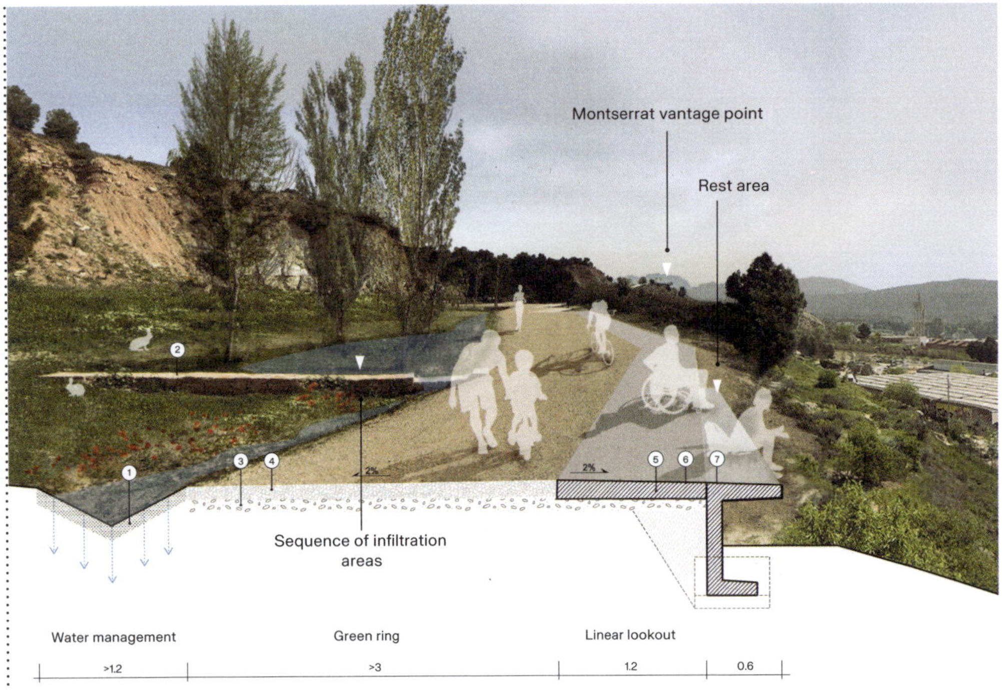

1 / Green drainage ditch
2 / Reused stone retaining walls
3 / Milled asphalt base
4 / Path of gravel and coarse sand
5 / Inclusive strip of concrete pavement
6 / Surface treatment of concrete with luminescent sand and gravel aggregate
7 / Precast concrete element for linear benches and spillways.

Type section of the path

The central strip, which is the main sandstone path, guarantees the continuity of the Green Ring, approximately three meters wide, and is arranged mostly according to the pre-existing layout of the area. An inclusive strip, in the form of a 1.2-meter concrete strip, complements the accessibility of the entire route and is located along the edge of the path.

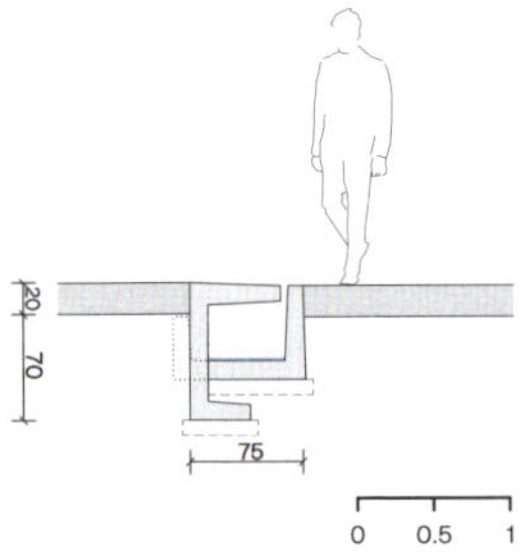

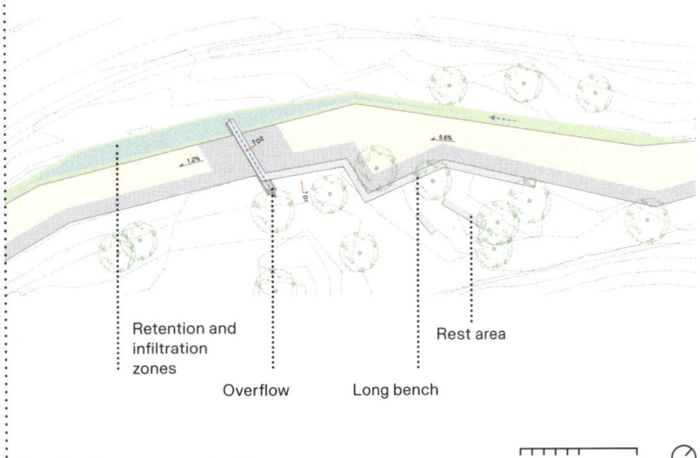

Details of rest areas and spillways

The Montserrat vantage point during the European Balloon Festival

The layout of the path takes in the main views of the city and its surroundings, integrates existing vegetation and trees, widening certain areas and creating rest areas along the route. It culminates in a cantilevered lookout on the slope of the mountain that offers spectacular views of the city and Montserrat mountain.

The decision was made not to have any street lighting, to avoid light pollution and considerably reduce building costs. The concrete pavement incorporates a luminescent aggregate, NIGHTWAY® STONE, that captures light during the day to signal the route at night.

Client Igualada City Council / **Authors** Enric Batlle, Joan Roig, Iván Sánchez / **Team** Mario Súñer, Yago Cavaller, Simone Cicu, Elisabeth Torregrossa / **Collaborators** STATIC – Structural Engineering / **Contractor** Moix Serveis i Obres / **Project area** 800 m

WELL Buildings, the User at the Center

Our buildings must achieve a degree of physical and material spatial comfort that promotes users' health and well-being. We must follow the "LEED is not enough" criteria by designing buildings and environments that, regardless of the certifications they are awarded, offer their users better living conditions, even in pandemic times.

In many residential development projects, we seek to increase the size of terraces, exceeding the classic Barcelona 1.5-meter-wide space, as in the **Infinitum Residential Buildings in Salou**. Here the dwellings intermingle with the existing pine forest, creating communal spaces, private gardens on the ground floor and numerous outdoor patios.

Prioritizing users' health and well-being

We need to design spaces that create comfort for users, given that we spend more than 90% of our time inside buildings. We have to implement design strategies that help improve users' morale and health, and even affect concentration, reduce absenteeism and depression, retain talent, improve user satisfaction and, at the same time, increase productivity.

Nature is always the answer

Our designs have to go beyond regulations and, with our projects, create new policies that update the obsolete restrictions we so often run up against. The pandemic made clear the need for direct contact with nature. We need buildings that have large outdoor spaces that their occupants, whether at home or at places of work or study, can use extensively.

WELL: air, water, nutrition, light, fitness, comfort and mind

We need comprehensive interdisciplinary approaches to effectively address the complex problems of human health and well-being. The WELL Building Standard® seal is based on a wide range of scientific disciplines. It takes a comprehensive approach that reinvents the built environment around its inhabitants, transforming the places where we live, work and learn into systems that promote and improve human health and well-being.

Generalitat de Catalunya Administrative District in Barcelona (Spain, 2016-2021): a complex designed to provide high-quality workspaces in connection with nature that promotes the well-being of people.

60% reduction in H_2O (14 l person/day) = 4 Olympic-size swimming pools

MWh/year savings = **284 tn** of carbon

CO_2 savings = **733** cars

1st building in Catalunya

LEED PLATINUM Certification

WELL GOLD Certification

The creation of this administrative district is the largest multi-departmental and corporate administrative concentration project ever promoted by the Generalitat de Catalunya regional government.
The complex, located in the Marina del Port neighborhood, was designed to provide quality workspaces. It is made up of two buildings linked by a service base, with a central landscaped area. This two-level plinth includes, among other things, the entrances to the two buildings, access to the communication centers of both, a customer service area, an assembly hall and a press room, in addition to numerous meeting and training rooms of different sizes and characteristics designed to serve the entire complex. The cafeteria and dining area is located on the top floor of this plinth, which allows for direct connection to the outdoor space of the campus.
Each 3,315 m² floor has a central area with two communication shafts around the courtyard that serve two symmetrical work wings between them. The floor plans are designed as open-plan spaces, generating work areas that are visually separated by the modulation of axes between pillars. However, the continuity of the workspaces is a function of the façade, which provides them with high quality lighting.

Type floor plan

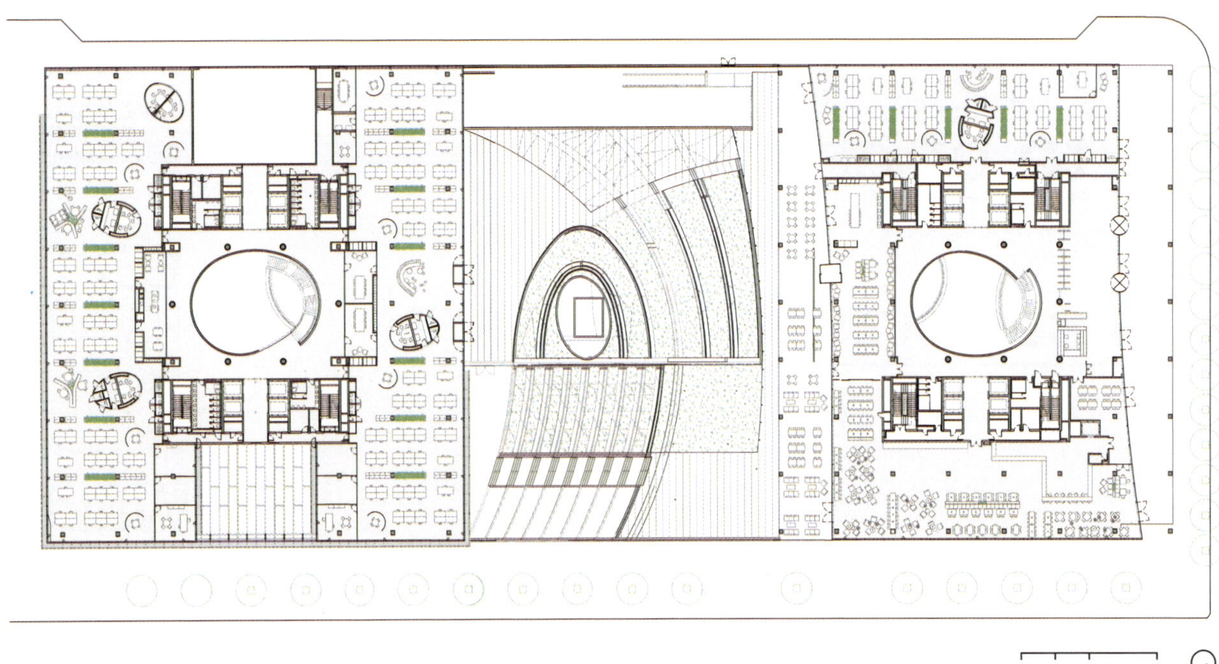

Ground floor plan

Sustainability and energy efficiency
The double-skin façade adapts to the building's four orientations to improve user comfort and promote energy efficiency, maximizing the visual relationship with the exterior and aligning with circadian rhythms.

Building

Recycling the existing foundation has resulted in lower CO_2 emissions than tearing it down and rebuilding it.

Promoting biodiversity by designing a biophilic building where vegetation is the protagonist on the campus, courtyards and interior spaces, threreby improving people's health, air quality and biodiversity in the city.

Promoting the circular economy by recycling and reusing the building's rainwater and gray water. Rainwater is collected from the roof and gray water from the restrooms, stored in tanks in the basement and used to water the building's vegetation.

Promoting sustainable mobility with bicycle parking at the main entrances to the building and in the basement. The project complements the city's new sustainable mobility network while promoting healthy indoor routes by including a new staircase in the central courtyard that is visible from all spaces.

Mitigating the heat island effect in the city by placing strategic elements on the roof such as vegetation and components with a high solar reflectance index.

Using local, environmentally friendly and recyclable materials.

Developing passive strategies to achieve minimum energy demand such as double-skin façades optimized by orientation.

Boosting the building's energy self-sufficiency and productivity by connecting it to the Ecoenergies grid and installing a photovoltaic pergola on the roof.

Placing the health and well-being of users at the center of the design strategy by promoting actions that improve spatial, physical and material comfort, such as the incorporation of the central campus, biophilia and use of natural materials.

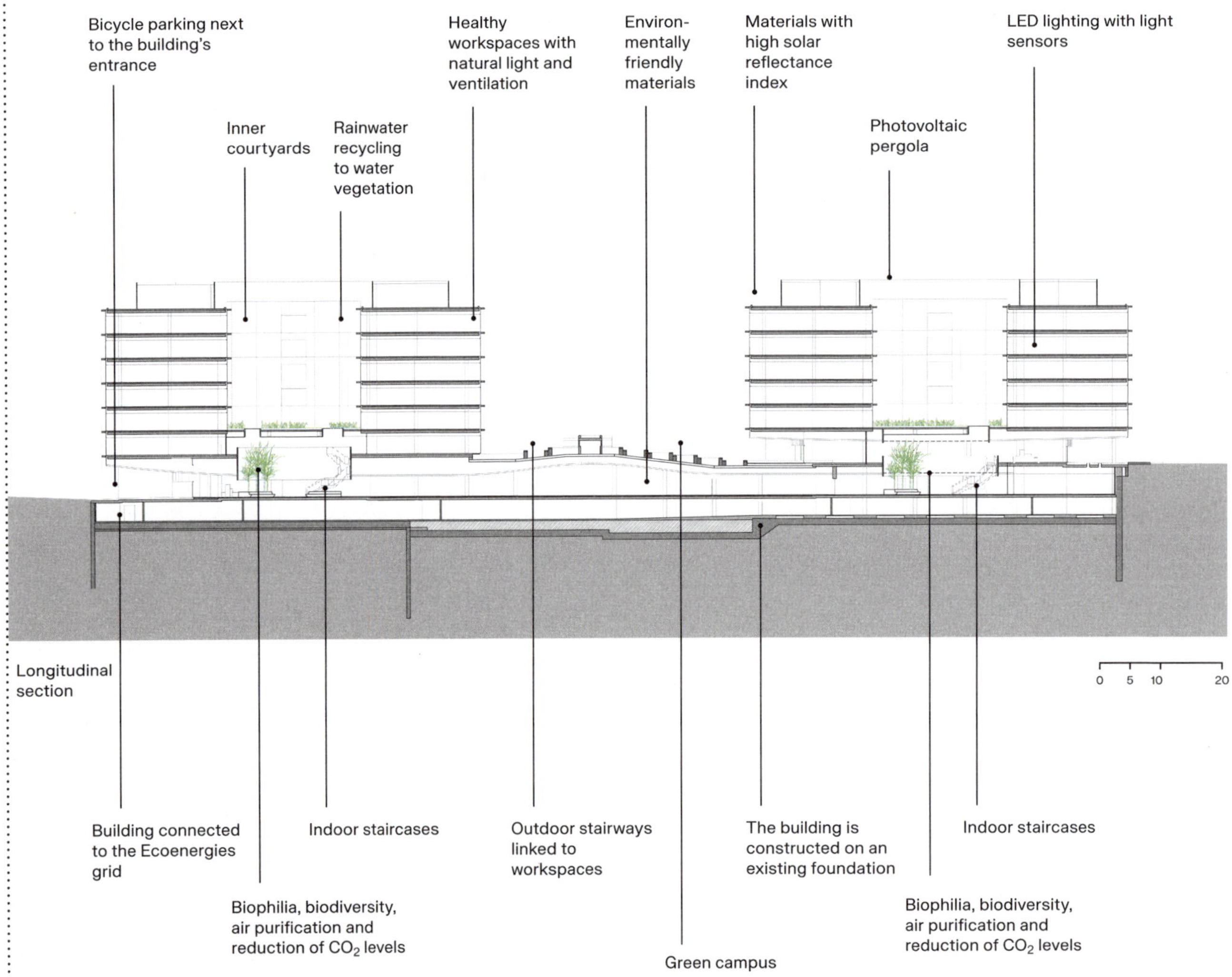

Longitudinal section

Building

The user at the center
Users can enjoy collaborative spaces close to workstations, which improve circulation flow and promote communication, characterized by biophilia that influences well-being and positive productivity.

The complex introduces ways of working and interacting through a new conception of spaces to achieve the objectives of maximum architectural and energy efficiency. User well-being is promoted through the materials used, the spatial layout and vegetation to generate biophilic spaces.

Client AXA Investment Managers, Hines, Infraestructures de la Generalitat de Catalunya / **Authors** Enric Batlle, Joan Roig, Albert Gil / **Team** Anna Lloret, José Sanz, Maria Comas, Pau Morgado, Paula Lladonosa, Maria Duran, Toni Duran / **Collaborators** BIS structures – Structural Engineering, PGI Engineering – Facilities Engineering, ARCO – Façade Consulting, G3 – Building engineering and UTE Fahe-Cba-JC – Executive Management / **Contractors** Comsa, Alainsa, Elecnor, Thyssen, FCC / **Gross floor area** 58,587 m²

The 3,000 m² outdoor space between the two structures was designed as a garden that functions as a campus and promotes social interaction in a healthy environment while providing better environmental conditions in this part of the city.

“The garden of the metropolis”, the title of Enric’s doctoral thesis, perfectly defines this attitude. Going beyond the classic concepts of urbanism, architecture and landscape, it seeks to connect them not only with landscape, but also with public space, prioritizing it and making it the driving force behind the construction of the city. This attitude has become the *modus operandi* of our firm, which always seeks to link the definition of the urban to different landscape logics.

A personal story

Joan Roig

Batlleiroig was born across a table in the studio of José Antonio Martínez Lapeña and Elías Torres, in mid-1981. Although Enric and I pursued our undergraduate studies on a parallel course, our paths did not cross until senior year, when Rafael Moneo and Elías Torres were our professors. Moneo had brought the ETSAB a scientific outlook and an inclusive attitude in contrast to the more restrictive Postmodern or *Tendenza* schools of thought. With him, we discovered the importance of place in the project, and the value of craftsmanship and professionalism in our work.

The fact that Enric's father was a gardener had a strong influence on our interest in landscape projects. This interest coincided with a time in Catalonia when public space competitions were beginning to be organized. While in the Barcelona of the sixties and seventies, the criteria for designing parks and gardens focused on adapting to the Mediterranean climate the model Burle Marx had generalized in South America, in the eighties the approach to public spaces followed the tradition of the Italian *piazza*. Our position at the time was instinctively the opposite. When it seemed that landscape design had to proceed from architectural elements, we chose to work with the classical elements of the landscape: vegetation, water and topography, and from a landscape perspective. Our first project, "Il Segretto Giardino", was submitted to the competition in 1981 for the Parc de l'Escorxador, a large park with historical features enclosed within four walls that isolated it from the city. Our second project, the Roques Blanques Cemetery, consisted of agricultural terraces in which the niches were housed in the retaining walls of the upper terrace protected by pergolas. Both projects, carried out in collaboration with Olga Tarrasó, eliminated architecture to become gardens or agricultural landscapes.

The fact that my father was a bricklayer did not determine my becoming an architect, but it did influence our studio's interest in construction. Alongside our landscape work, from the outset we had a fervent desire to build, spurred by the boom in public competitions in the eighties. Not specializing in a specific typology allowed us to take on construction projects ranging from public housing to schools, markets, libraries and even bridges. Because we were willing to work on all kinds of projects and at all scales, from urban projects

Park in the Trinitat road junction, Barcelona (Spain, 1992)

to more specific design endeavors, for us the eighties were an incredible learning experience. Our ambitious standards, the importance we gave to drawing when defining the project and the professionalism with which we have always approached the management of construction work allowed our team to increase in number and qualifications. The construction of Nus de la Trinitat Park, for the 1992 Barcelona Olympic Games, signaled the start of a process of designing large parks linked to city projects, as in our work in Sant Cugat, Viladecans, Zaragoza and Santander; or to territorial projects, such as the Garraf landfill. This led to the addition to our team of the role of agricultural engineer, the seed of our current landscape department.

"The Garden of the Metropolis", the title of Enric's doctoral thesis, perfectly defines this attitude. Going beyond the classic concepts of urbanism, architecture and landscape, it seeks to connect them not only with landscape, but also with public space, prioritizing it and making it the driving force behind the construction of the city. This attitude has become the *modus operandi* of our firm, which always seeks to link the definition of what is urban to different landscape logics.

The economic crisis in 2008 prompted a rethinking of policies concerning large parks, leading to a less generalist but more structured model. In this sense, our current work with large road constructions to organize the territory, such as the Llobregat Blue Route or the cycling connections between Barcelona and its metropolitan area, is a clear example of this shift. Reusing landscape in a new structure of movement encourages a paradigm change. In terms of buildings, on the other hand, while the crisis brought the system of public commissions nearly to a halt, in our case it had a revitalizing effect in the private sector. We had previous experience in office buildings, such as the INTERFACE building and CMT headquarters in Barcelona's 22@ district, but as of 2008 we became extremely active in this area. This allowed us to study different cases in relation to the structure, enclosures and environmental performance of this type of building. Meanwhile, thanks to our relationship with INDITEX, we began to work on large buildings devoted to the coordination of major fashion brands. The paradigm shift brought about by the crisis led us to train all our teams in the

Atlantic Park in La Vaguada de las Llamas, Santander (Spain, 2008)

use of new technological tools such as BIM, and to incorporate sustainability, the environment, well-being and health as cornerstones of our design process.

Since our beginnings we have tried to be specialists in each of the disciplines we practice (City and Territory, Landscape and Public Space, Architecture and Construction), never ceasing to innovate and never forgetting the essential cross-cutting approach needed to develop any project. We view urban planning, landscape architecture and architecture as a complex process in which various creative activities are superimposed, using different techniques to obtain maximum efficiency in the results. The exponential growth in the volume of commissions in the last decade and their increasing specificity has led us to reorganize Batlleiroig by means of a new corporate governance system. Two new partners joined the firm in 2014 and four more in 2020. Currently, Batlleiroig has eight partners, several associates and over 140 professionals. Together, we form a solid team that is always learning and ready to continue offering its best in our commitment to the future.

Corporate headquarters of the Telecommunications Market Commission, Barcelona (Spain, 2010)

Former Mercedes-Benz factory, Barcelona, June 2, 2022.
The entire Batlleiroig team at the party to celebrate the firm's 40th anniversary.

batlleiroig
team

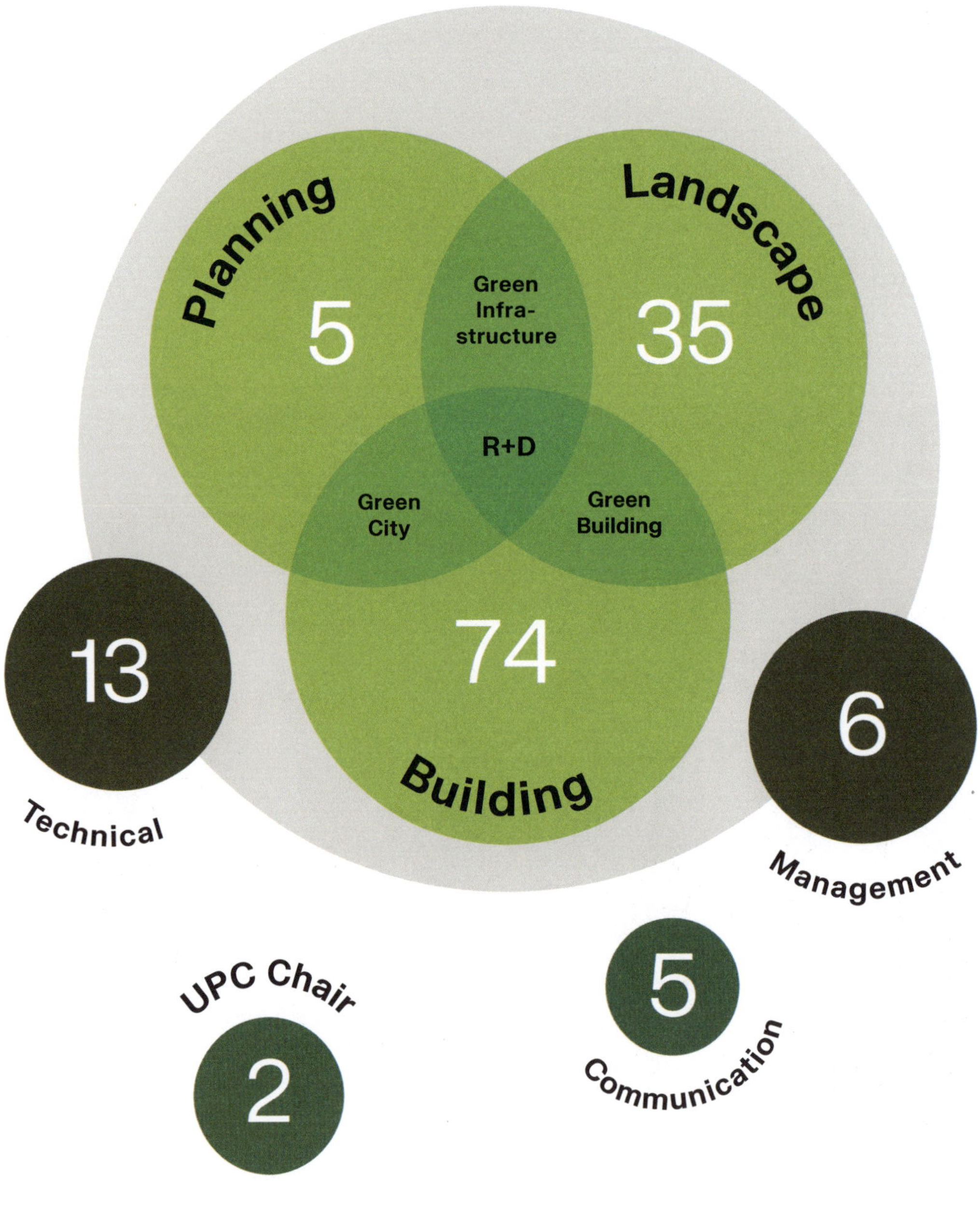

Enric Batlle Durany ×+
Founding Partner
Doctor of Architecture & Landscape Architect

Joan Roig i Duran ×
Founding Partner
Architect

Albert Gil Margalef ×
Senior Partner & CEO
Architect

Planning

Abel Porcar Badal ×◆
Director & Partner

Helena Valls Fígols ×◆

Ignasi Rincón Riquelme ×◆

Mercedes Blay Blanc ×◆

Tomás Barreiro Vescovo ×◆

Landscape

Ivan Sánchez Fabra ×
Director and Senior Partner

Joan Batlle Blay ×+
Vice Director and Partner

Mario Súñer Díaz ×
Vice Director and Associate

Antoni Monté Llopis ×★
Associate

Albert Nolla Parals ×
Bárbara Freire Ares ×+
Begoña Roget de Aysa ×
Carles Esquerra Julià ×
Clàudia Amías Roget ×+
Cristina Casanovas López-Amor ×
David Bravo Villafranca ×
David Pérez Cuello ×
Francesc Montero Llorca ×★
Gabriel Romero Fossas ×
Gemma Montané Cartró ×
Héctor Castillo Barrio ×
Isabel Pérez Manen ×
Jordi Juanola Rosselló ×
Jorge Espinosa Morales ×+
Laura Florentín Aragón ×
Lidia Carrillo Parra ×+
Marc Ruiz Llorente ×
María Moure Rosende ×
Marina Santos Loshuertos ×
Mario Monclús Bazanbide ×
Marta Ines Cot ×+
Marta Sanz Cuso ×
Mauricio Morfín Quintanar ×
Mercè Lorente i Gras ×
Òscar Torra Moneny ×
Patricia Flores Hernández ×+
Roger Mermi Ribalta ×
Violeta Garcia Torralbo ×
Xavier Llerena i Bastida ×

—

Santiago Cavaller Galí ▲
Biodiversity Director

Building

Helena Salvadó Giné ×
Director and Partner

Meritxell Moyà Vilalta ×+
Vice Director and Associate

Anna Lloret Papaseit ×
Vice Director and Associate

Josep Batlle Blay ×
Director and Partner

Beatriz Barral Santos ×
Associate

David Sesé López ×
Associate

Laura Quintana Fernàndez ×
Associate

María Lourdes Romeo Sus ×■
Associate

Adrià Loza Gutiérrez ×
Ainhoa Fernández Calleja ×
Alba González Vilar ×
Alba Moreno Rodríguez ×
Aleix Masjoan Colomer ×
Ana Gómez Fernández ×
Ana Valverde Cabello de los Cobos ×
Anna Porta Almagro ×
Annabel Barba Calpe ×
Begoña de Pablos Alfaro ×
Blanca Hinojosa Arner ×
Clara Caldes Ortín ×
Clara Doyhambehere ×
Claudia Márquez Sala ×
Claudia Senovilla Martínez ×
Cristina Batalla Vilacis ×
Cristòfol Tauler Bennàsar ×
David Balseiro Garcia ×
David Casado Pérez ×
David Guzman Lirio ×
Diana Martínez Badía ×
Edurne Oyanguren Zorita ×
Eirene Presmanes Yrurzun ×
Elena Coronas Soucheiron ×
Elena Miralles Coll ×
Esther Díaz Salas ×
Fernando Marí Rebollo ×
Francisco Javier Luna Gómez ×
Irene Ayala Castro ×
Irene Escudé Moreno ×
Javier Blasco Claveria ×
Javier E. Castillo Sánchez ×
Joan Balle Llabrés ×
Joane Amasorrain Larrañaga ×
Jordi Llàcer Macau ×
José Sanz Gorordo ×
Josep Maria Navarrete ×
Josu Acebrón Gutierrez ×
Laura Menéndez Smith ×
Lucía Ortiz de Galisteo López ×
Manuel Pedraz Salas ×
Marc Sánchez Olivares ×
Maria Comas Giménez ×□
María Ferrer Morey ×
María Méndez Gago ×
Mariano Aguado de Sas ×
Marina del Aguila Ramos ×□
Marta Gil Margalef ×
Marta Gragera Caballero ×
Martin Hauge ×
Miquel Manén Gorgas ×
Nerea Castell Sagües ×
Nieves Gómez García ×
Núria Ribas Ortiz ×
Pablo Macías Trujillo ×
Pau Morgado Martí ×
Pau Solés Gimbernat ×
Paula Lladonosa Farré ×□
Paula Ortiz Arias ×
Pilar Álvarez Alvargonzález ×
Rafel Bonet Sisniega ×
Rita Pérez Morandeira ×
Ross Pérez Aiyer ×
Santi Xercavins Julià ×
Margaret Tara Maalouf ×
Xabier López Corres ×

Technical

Diana Calicó Soler ●+
Associate and Director of Building Engineering

Carolina Valero Ramos ●
David Sobrino Viguín ●
Elisabeth Torregrosa Avilés ●
Gemma Ramon Mestre ●
Helena Canals Albiol ●
Jordi Pontnou Orga ●
Laura Luque Pan ●
Victor Susagna Holgado ●
Marc Torrella Mayol ●

—

Antonio Duran Bautista ×
BIM Director

Agnès Parcerisa Giné ×

—

Alberto Blay Blanc
IT Director

Management

Vanesa López Blanco
Director

Anna Gracia Moegle
Elisabet Bueno Sanchez
Miriam Gómez Ortega
Selena Gómez Gámez
Vanessa Maneiro García

Communication

Ana Badía Alzuria ×
Director

Alicia de las Heras Benito ×
Inés Piñeiro Ozores ×
Paula Amargós Molina ×
Paula Piñeiro Ces

× Architect
+ Landscape architect
◆ Urban planner
● Building engineer
▲ Agricultural engineer
□ Interior design expertise
★ Quality management expertise
+ Façade expertise
■ Sustainability expertise

Merging City and Nature
30 Commitments to Combat Climate Change

Published by
Actar Publishers, New York, Barcelona
www.actar.com

Authors
Batlleiroig

Edited by
Marta Poch

Actar coordination
Ricardo Devesa

Batlleiroig coordination
Ana Badía

Graphic design
spread. Tomoko Sakamoto - David Lorente

With contributions by
Enric Batlle
Joan Batlle
Joan Busquets
Clara Olóriz Sanjuán
Joan Roig
Marta Thorne

English Translation
Kevin Alan Krell

Proofreading
Elaine Fradley

Typeface
FK Grotesk Neue BIR by Florian Karsten

Printing and binding
Arlequín & Pierrot SL

ISBN: 978-1-63840-009-7

Printed in Spain
Date of publication:
November 2022

Available in Spanish
Fusionando ciudad y naturaleza: 30 compromisos para combatir el cambio climático
ISBN (Spanish): 978-1-63840-010-3

Cover: Pedestrian Connection Between Barcelona and Montcada i Reixac, 2019-2020 © Jordi Surroca

The publisher has made every effort to contact and acknowledge copyright holders. In the event of any omission, please contact the publisher, who will make necessary changes in subsequent editions.

Photographic credits

© Antonio Navarro Wijkmark, 31, 62, 63, 64, 65, 128, 130, 131, 154, 156, 157, 158, 159, 160-161, 194, 197 (left), 198, 199, 210, 212 (btm), 213, 266, 269, 270, 271, 272, 283, 294, 302, 303, 304, 305, 323 (btm), 336, 339, 341, 342, 343, 344-345, 349, 354-355
© Archivo DOCOMOMO Ibérico, 317
© Arxiu Municipal de Sant Fruitós de Bages, 204 (left)
© Ariel Ramírez 240, 244
© BIR, 13, 16 (right), 54, 60, 73 (mid.), 82, 84, 88, 81, 90, 91 (btm), 92, 103, 116, 118 (btm), 121, 124, 126, 166, 178 (top), 197 (right), 200, 204 (right), 212 (top), 234 (btm), 261 (btm), 284
© DEL RIO BANI, 334
© Goa Invest, 300
© Google Earth, 252
© Google Maps, 174
© Jordi Miralles, 182
© Jordi Surroca, 6, 15, 19, 21, 24, 25, 33 (top), 43, 44, 45, 46, 49 (top), 47, 48 (top), 50, 52, 55, 56, 57, 70, 73 (top/btm), 74-75, 77, 79, 80, 91 (top), 93, 108, 111, 112, 113, 118 (top), 119, 122-123, 136, 138, 139, 140, 141, 142-143, 146, 149, 150, 151, 168, 170 (top), 176, 178 (btm) , 179, 180, 181, 184, 187, 188, 189 (btm), 202, 205, 206-207, 230, 232, 233, 234 (btm), 235, 236, 237, 256, 258, 260 (top), 261 (top), 262-263, 286, 288, 289, 290, 291, 292-293, 326, 328, 329, 330, 331, 332 (btm), 333, 351, 353, 358
© Jorge Poo, 106
© Institut Cartogràfic i Geològic de Catalunya, 46 (btm), 48 (btm)
© Luis Díaz Díaz, 33 (btm), 296, 301, 306, 307, 308, 309
© Oriol Gómez, 264
© Quim Bosch, 16 (left), 228, 260 (btm), 332 (top)
© Rafael Vargas, 96, 99, 100-101, 102, 103 (right), 278, 280, 281,
© SBDA, 23, 32, 58, 94, 152, 170 (btm), 189 (top), 192, 208, 216, 217, 222, 225, 226, 227, 238, 242, 250, 253, 254, 255, 282, 298-299, 312, 314, 316, 318, 320, 321, 323 (top), 324
© Stupendastic, 171, 172-173, 245

Distribution
Actar D, Inc. New York, Barcelona.

New York
440 Park Avenue South, 17th Floor
New York, NY 10016, USA
T +1 212 966 2207
salesnewyork@actar-d.com

Barcelona
Roca i Batlle 2-4
08023 Barcelona, Spain
T +34 933 282 183
eurosales@actar-d.com